Sex Robots

Sex Robots

Jason Lee

Sex Robots

The Future of Desire

Jason Lee
Leicester Media School
De Montfort University
Leicester, United Kingdom

ISBN 978-3-319-49321-3   ISBN 978-3-319-49322-0 (eBook)
DOI 10.1007/978-3-319-49322-0

Library of Congress Control Number: 2016957756

Cover illustration: Mono Circles © John Rawsterne/patternhead.com

Printed on acid-free paper

This Palgrave Macmillan imprint is published by Springer Nature
The registered company is Springer International Publishing AG
The registered company address is: Gewerbestrasse 11, 6330 Cham, Switzerland

CONTENTS

Contents

Robotic Evolution

Abstract Chapter 1 examines the myths, legends and history of robots, emphasising how the desire for robots has always been with us. We see where these desires for sex robots stemmed from, and what they mean in a global context. Starting with the Greeks, and moving up to the present day, the importance of Margaret Atwood's work, especially her novel *The Heart Goes Last* is stressed. Many examples are examined, from children's books and films, to works of philosophy. Using a variety of thinkers, such as theologian Karen Armstrong, this opening chapter explains where the fears connected to the sex robot stem from.

Keywords Sex robots · Sex · Robots · Myths · Legends · History · Desire · Greek · Margaret Atwood · Karen Armstrong · Philosophy · Theology · Children · Pygmalion · Golem · Android · Automata · Mannequin · Heidegger · Baudrillard · Fetishisation · Gratification · Fertilisation · Postmodernism · Frankenstein

Sex robots – or sexbots as they are sometimes called – are constantly cropping up in contemporary media discourse. Clearly, the sex robot symbolises far more than a human-made object to have sex with. Questions concerning difference and otherness, known as alterity, are central to discussions over sex robots. The sex robot challenges what it means to be human and simultaneously enables us to reflect on human nature itself. So, are we in the age of

© The Author(s) 2017
J. Lee, *Sex Robots*,
DOI 10.1007/978-3-319-49322-0_1

the sexbot? Hopefully, this brief introduction and the related chapters look-ing at science fiction and fact show that this phenomenon is not new, having been within myths and storytelling for centuries. A follow-up question is: should we be overly concerned? There is a strong argument, proposed here, that the sex robot has been with us in our dreams, genes and desires, since our origins, and is actually a benign influence. This book explores some of the myths surrounding the sex robot, its philosophical and cultural implica-tions, its evolution as a form historically through culture, and the scientific position we are now in.

Globally, myths connected to robots go back to ancient times. Even the earliest origin myths contain references to animated life being made from inanimate matter. The sex robot has been at the heart of the human imagination ever since culture as we now know it began. An obvious pre-eminent example is the famous myth of Pygmalion, whose statue of Galatea came to life. In the Greek story, Aphrodite grants Pygmalion's wish and he marries the ivory sculpture that has metamorphosed into a woman, and they have a daughter Paphos. The myth of Pygmalion, who falls in love with one of his sculptures, had a huge influence on later culture, such as Victorian writers. It was the title of George Bernard Shaw's play, which premiered on 16 October 1913 at the Hofburg Theatre in Vienna. Pygmalion is most widely known because of the account in Ovid's narrative poem *Metamorphoses* published in eight AD, the source of Ovid's work being a second-century myth. The impact of this work cannot be overstated. Ovid was translated into English in 1480, and has inspired many writers, such as Dante, Boccaccio, Chaucer and Shakespeare, plus numerous artists, includ-ing Titian, Michelangelo, Picasso, Rembrandt, Raphael and Rubens.

As well as Pygmalion, we also have in mythology the talking mechanical handmaidens built out of gold by the Greek god Hephaestus, known as Vulcan to the Romans. The re-animated clay golems of Jewish and Norse legends all relate to inanimate matter gaining life, and there is the Chinese legend of Yan Shi who made a human-like automaton, which is narrated in the *Liezi*. Legends of these robot forms exist in all cultures, frequently with an underlying element concerned with relationships and sex. These include many Indian and Egyptian legends, plus the famous Christian legend of Albertus Magnus, who in legend constructed an android to do domestic tasks. Of course, such myths relate to our deep desires to be free from the burden of work, but then they add to further fears of these robots actually replacing us and making us superfluous, like the child replacing the parents, as we shall see.

The thirteenth-century philosopher and Franciscan friar Roger Bacon was believed to be the creator of a metal speaking head; these legends began taking hold and having wider credence from the sixteenth century. As noted, it can be observed that globally there has always been a desire for some form of robot. Legend and myth often fused with storytelling to produce a version of the truth where such robots were possible. As Rob Pope (2005) explains, with reference to Giambattista Vico, all developments of the human through self-creation are 'expressed and apprehended through metaphor and symbol'. Vico, publishing *La Scienza Nuova* in 1725, proposed a science of man, a form of physics of the mind, given man had actually created himself. We shall return to science fact in our final chapter, examining current developments, but it should be clear by now that the first science was actually mythology and the interpretation of fables.

In 1769 Wolfgang von Kempelen, a servant of Maria Teresa, the empress of Austria-Hungary, witnessed magic tricks in Vienna and vouched he could do better. Six months later, in 1770, he appeared before the empress with a life-size mannequin, which became known as 'The Turk'. More was written on this automaton than on any other. This chess-playing machine could outplay most opponents, and toured Europe, with many people conjecturing on how it could work but not finding the answer. After Kempelen's death, the Turk was bought by Johann Maelzel, a maker of musicical automata, who invented the metronome. The Turk continued touring, playing chess with Napoléon Bonaparte in 1809, and twice beating Charles Babbage, the pioneer of the mechanical computer. Babbage concluded that the Turk must have been controlled by a human, and even though he could not work out how it worked, it made him want to strive to build a genuine machine. A twenty six-year-old Edgar Allan Poe came across the Turk in 1835 in Richmond, Virginia, and he believed there was a hidden operator. While the Turk was a hoax, an array of quite complex mechanical creatures were being exhibited across Europe at the time, which led to the belief in the machine being genuine (Standage 2002). The Turk was destroyed in a fire in a museum in Philadelphia on 5 July 1854.

The Industrial Revolution, starting in Britain, and then spreading throughout the world, commonly dates from the middle of the eighteenth century to the middle of the nineteenth century. This witnessed the development of the predominance of machinery, and with this came notable stories concerning the fear of this dominance. In 1898 H.G. Wells published *The War of the Worlds*, which was subsequently adapted in numerous ways, such as the 1938 real-time broadcast by Orson Welles which fooled many into believing an alien robot invasion was actually happening. Once again, however, we need to realise the symbolic meaning of this story. The Martians invading the earth can be compared to colonial empires already on earth, like the Victorian Empire, dominating other countries. Questions raised by this work are over what the robots will do with humans; will we be their pets and slaves? If vice versa, and we turn robots into our slaves, what are the ethics of this, a question raised by many films.

By the twentieth century the stories, myths and legends were becoming even more of a reality, at least in terms of entertainment. In 1928, Gakutensoku, Japan's first robot was designed by biologist Makoto Nishimura. Japan is still leading the world in terms of production and demand for robots. Robots then appeared at the World's Fair in 1939, including Elektro, who was apparently able to speak 700 words, smoke cigarettes and move. These robots were far from independent, however, and were often merely mechanical magic tricks like the Turk, for gullible sightseers. They functioned by containing gears operated by humans, but they offered a taste of things to come.

RealDolls have been producing sex dolls from the late 1990s. Strictly speaking, these are not robots but sometimes there is no difference between these functional dolls and robots in terms of what they signify and represent. Philosophically, what is known as the 'uncanny valley', when the sex doll or robot is just too like the human producing a revulsion, means many of the questions raised by these various forms are identical. Paradoxically, there is then an attempt to make these dolls less human, a point some science fiction writers have not absorbed. Part of the moral outrage over all of this concerns the belief that we are transgressing some form of sacred boundary, or at least challenging the human. The argument is that if we treat robots like objects then we objectify each other more, which then damages relationships. As we shall see, the converse could just as equally be argued.

The company RealDolls does have their boundaries. For example, they refused to make a sex dog, despite being offered $50,000, 1000 per cent higher than their normal prices. There are hermaphrodite versions, given some clearly want a vagina and a penis. There is a whole plant making male versions, and overall, in terms of female versions, there are 11 body types, with 31 faces, plus 30 styles. Customers select shades of nipples; skin and lip type; hair and eye colour; pubic hair (trimmed, natural, full, shaved); eyebrows (fake, human hair); removable tongues, tattoos, piercings; oral inserts (such as the 'deep throat'). Again, there are boundaries; the company will only make copies of celebrities if they get permission. Typically, only ten percent of customers are female, although this does not mean that the buyer then keeps the product to themselves. Perhaps the male purchaser shares their sex robot, doll or toy with their partner, be the partner male, female, transgender or beyond gender. There is also the opportunity to 'become' a sex doll, with equipment which includes built-in genitalia (Gurley 2015). We shall explore further the nineteenth-century fear of the human becoming the machine, but the owners of RealDoll do not aim to replace organic forms, just enhance the experience.

In the mechanised culture, all desire stems from a recognised element. Robot and sex that is sex robots, or sexbots, are therefore the natural equivalent within society. There is no separation between sex and robots, or between nature and robots. The ontology of Being and being within nature, following Heidegger, and the ontology of Being and being within culture, are synchronous. The sex robot, without equivocation, is in fact the quintessence of Being and being. Indeed, commonly and innately the machine in the human is more human than the human. There is no need for concern, however. As many of our fictions show, some explored here, these are not more than human, as some transhumanists would wish for.

Pornography, as it is commonly put, continually contains a robotic element. This is neither a condemnation of pornography, nor a defence, regardless of the medium and form. Whether this is morally conducive to a healthy sense of being is immaterial. The mechanised process of image reduplication, which renders an essential being immaterial, means that only the machine can be recognised as natural. To not have a relationship with the sex robot in this context would be immoral. The sex robot is not a mirror of the sexual human, or merely the object that is desired. Nor is the sex robot and sex with robots the quintessence of narcissism, which is a common misperception. The sex robot, put simply, is the transcendence of all being, moving beyond the ephemeral subject-object divide, the difference in the difference. In this sense, we move beyond Jean Baudrillard's celebration of indifference, or even his tautological arguments concerning 'only the other knows' (Baudrillard 1990).

What is natural or unnatural might actually not be a question at all. The sex robot can go to places that the human cannot. Moving beyond the limitations of the cellular form, and the underdeveloped uninvolved, paradoxically existing only naturally in and of its own self. If we need to bring God into the equation, we can confirm that, within the theology of human development, this is a quintessential stage, and to deny it would be to deny our own being. Logically, if there is no distain in and through nature, and the sex robot is an intrinsic part of nature, then there is no disdain in and for the sex robot, or with any interaction concerning the sex robot. The sex robot is the manifestation of our deepest desire, and therefore is our dream manifestation. We become ourselves when we become our dreams therefore becoming our sex robot is the goal. 'Get up (I Feel Like Being a) Sex machine', to quote James Brown. Within this July 1970 song there is the common desire to move beyond the human, and so to step into a sign that is transcended. The transcendence of the machine and a utopia that comes from this transcendence has a long history. The sexual element of all that is mechanical stems from the desire of the human. Within certain eastern discourse, being and Being is an illusion, as is thought, including the limit of thought itself. The mind is a mental continuum within this paradigm, and the body might be deceptive. All desires need to be removed according to these philosophies, but the body links us to duration and time. This then, paradoxically, brings us to the real world.

We shall see how predetermined the truth of desire is, which paradoxically offers a creative freedom, and its manifestation as the sex robot as truth is paradigmatically obvious. In Margaret Atwood's 2015 dystopian novel, *The Heart Goes Last*, there are different levels of sex robots, some that actually speak and respond 'naturally'. These higher evolved species of robot can communicate. People have taken tests finding, fundamentally, on a performance level, that it is difficult to distinguish between a human and a robot. So, a sex machine, or sex robot, and a human having sex, are all identical, although we need to contest whether this should be the aim of sex robots. Perhaps the aim of pure reduplication is not ambitious enough. It is essential for us to see beyond our own paradigms.

Dildos that are robotic and vaginas that are robotic are not the same as sex robots, of course, although remote controlled sex aids that can be used internationally are becoming mainstream. You might be lying in a hotel room in Japan and your partner is in Brazil, but through technologically advanced sex aids you can have 'sex'. What is wrong with the phone, or a pen, all forms of technology in themselves? Nothing; only why not use the technology that could be available, which actually brings you closer. I need to put 'sex' in this way, as it is clearly recognised that non-embodied sex is beyond traditional frameworks. We shall come to this point, however, when we explore fictional films that highlight the importance of the voice over the body, in terms of sex.

Overall, these aids subvert the view that these forms of technologies concern alienation and loneliness. In a globalised world where people, to quote a former Conservative politician who shall remain nameless, must 'get on their bike' to find work, anything that may enable relationships to continue over long distances should be praised. The usual other attack on the use and development of these machines comes from the position that this is about promoting misogyny, and the objectification of women, a subject we shall come to. But sex robots can be of any particular sex and sexuality.

What makes us human, and what makes us our gender, are tougher questions than we think provoking intense debate. In April 2016 novelist Ian McEwan apologised for saying he tended to think having a penis makes you man, then claiming gender is not biologically determined. Academic Germaine Greer has raised controversy for her opinion that having a sex change does not make you a woman. We can then theoretically ask ourselves what college or university may robots be able to study at, if the colleges are separated along gender lines, such as Bryn Mawr College for women. As Andrea Dworkin explains,

> 'man' and 'woman' are fictions, caricatures, cultural constructs. As models they are reductive, totalitarian, inappropriate to human becoming. As roles they are static, demeaning to the female, dead-ended for male and female both. (Dworkin 1974)

Then there is the notional idea of robots using toilets, even if this is just to mimic the process; which toilets would a sex robot use? In the children's film *Robots* (Chris Wedge 2005), understandably, sex is not particularly emphasised, although there are many jokes that concern sexually identified zones of the robot, such as the anus. We shall come on to the world of Roald Dahl, where similarly the threats of advanced robots to the family are made apparent, but then overcome. There is comedy and laughter concerning art, that may cause pleasure, and there are traditional family zones of laughter. But often the sexuality is denied, repressed or ignored.

It is as if the children are to be protected from any knowledge of sex when it comes to robots and machinery. This is despite the overt message of invention in the film *Robots*, and the undertone of fertility and creativity being at the heart of this story, and many other children's films. In *Inside Out* (Pete Docter and Ronnie Del Carmen 2015), for example, the feelings of a young girl are constructed as creatures inside her mind. These mechanistically control her memories, both good and bad, and the inner workings of her mind parallel that of a robot. They are solidified into balls, until her mind is a type of pinball game, also reminiscent of an executive toy. The Newtonian universe, where the planets and all matter run in a mechanical fashion, is internalised. Although, given this is a twentieth-first-century film, quantum theory is accounted for with seemingly random yet helpful events within the narrative. The film does not portray a girl in the sway of emotions, because free will is emphasised, but what is of note here is that the storytellers are using the idea of the mind functioning like a machine.

Asking whether you would have sex with a robot is assuming a robot would have sex with you. Just as we may question what it means to be human, we can question what sex actually is. One might assume a sex robot is pre-programmed to be accommodating to your sexual needs, but what if a sex robot was programmed to have its own individual preferences that overcome yours? There is a sublime element to the sex robot. Sexual activity for the human often concerns an event, with an outcome. Simply put, this can be gratification, reproduction through fertilisation, or any amount of outcomes, including confirmation of the ideological system. One argument that is not proposed enough is that the problem with sex robots is that they are all about consumerism. From the cradle to the grave you might be relying on this helper, be it for potty training, sex, or with final life care and this will cost you. Promoting the need for sex robots is the same as promoting the need for anything else, and as advertisers know, sex is normally at the heart of creating these needs.

With the robot, however, there is activity which may involve an outcome of energy transference, or it may not involve any of this. It involves repetition, perhaps, within this paradigm, and it involves algorithmic reduplication, one would assume. But it does not involve any connection that may involve death directly, given its invention and actual creation is a method to overcome death. Unlike the human occurrence, this fleeting moment of transcendence, if we are lucky, is not *le petit mort*. When it comes down to it, any robot challenges how we perceive identity. Even early cybernetics saw that what makes up identity is not the matter from which a physical identity is made. Perhaps all we can say is that continuity of processes forms an identity, but even this is arguable. Whether corporeal or incorporeal, organic or inorganic, or a fusion of various elements that moves us away from these limitations through robotics, we find a new version of the human (Davis 1998). The machine theoretically can never die and in this sense any orgasm is not le petit mort, but the reverse. It is a little acknowledgement of the eternal potential. This is the essence of utopia, the essence of all religion. There is hope within this sexual relationship with the machine that goes beyond any human limitation.

The fetishisation of the machine, that is so common within popular culture, such as cars within James Bond films, is just one small element of this sexuality. But what we see within that discourse most commonly is destruction, and a determination to terminate any lack of potential with cars and guns. They are usually discarded and then, after the moment of transgression, the human re-emphasised. Here then the state is re-enforced as the essential zone of protection, the mother figure, Judie Dench's incarnation of M. Sacrifices must be made. Paradoxically, the robot and the sex robot in particular, signify abundance. This signifies not *le petite mort*, but *le grande vie*. This in itself may be the threat to the ideological status apparatus because no longer is there the traditional requirement for sacrifice.

Importantly, the desire for the machine is equalled by the desire to be a machine. Bond here is yet again another example, attacked by his critics as a 'killing machine', but also 'licensed to thrill'. To be sure, being a machine can make one at one with the ideological system, but it can subvert this system. This we need to recognise, however hard it might seem, is not an either or scenario, our postmodernism position carrying us beyond such binaries. Being at one with anything is in itself a sexual act. There are a number of ways where the being can become a robot to survive, and so humanity becomes robotic. We do not mean the human metaphorically, but what we mean is an electronic transfer of matter to inorganic materials, which allows life to remain. In this sense, a sex robot, which is a transcended organic being, is the essence of organic.

This is no mere sophistry, but is real depth psychology, with the materialisation of a fusion of the canny and the uncanny psychic elements. As with the clone, who raises a question over authenticity, identity, and individual human significance, the well-made sex robot challenges all our preconceived conceptions of what is human and what makes us humans. There is an obsession with narrative, with creating our own story that developed out of the New Age movements of the 1970s, typified by the work of Sheldon Kopp (1983). It is easy to mock the Kopp-style movement, although Kopp's major point was that there are no hidden meanings in life and our lives have only the significance we give them. Thus, 'Killing the Buddha on the Road' means destroying the hope that anything outside ourselves can be our master. The laboratory aspect of this is obvious; you do not have to be a slave to the machine, or a machine for the system. An obsession for 'finding the real you' began during this period, and took off with celebrity culture, and the Oprahfication of America. Forget the problems over potential nuclear war, climate change and world poverty, the real issue was: who am I?

The sex robot was and is the ultimate threat to this pursuit of 'the original' because, like the clone, it can challenge the myth of uniqueness, so central to many belief systems, especially monotheistic. Thankfully, as ex-nun and pre-eminent theologian and scholar Karen Armstrong (2009) has shown most eloquently, the real religious experience is about transcendence. One would hope the best sex robots bring about this form of transcendence, at least minimally. When we view this experience as ultimately a mental phenomenon, regardless of its spiritual resonances and implications or its apparent physical causation, should we really be condemning how this experience of transcendence is caused?

As the title of Philip K. Dick's 1968 novels puts it, the question also arises, *Do Androids Dream of Electric Sheep?*, or do electric sheep dream of humans dreaming of androids; press repeat. The fact that androids dream of androids is no mere suggestion, but a factual explanation that the unconscious is a mechanically driven entity. We do not have to be hung-over with the nineteenth century, like Freud, to accept this. The unconscious of the human is in fact the conscious of the sex robot. The unconscious is always one step ahead of the conscious, and the future is the sex robot. To understand this is to understand the purpose of dreams, and the purpose of the unconscious. As analytical psychology proposed, we must be prepared to realise that we do not understand by going outwards, but by going inwards. That is where we find not the ghost within the machine but the machine within the ghost.

When the ghost in the human is the machine, and the machine is the ghost, it is clear that all that is paranormal and supranormal is the sex robot's desire and desiring. The desire then for this other is not external to the self. The sex robot is always driven by the aesthetic, and this is another challenge to the human that might turn the human against the sex robot. For example, we know that the Kantian subject when encountering beauty discovers in it harmonious relations that are in the manifestation of the unrestricted play of its own faculties (Eagleton 1986). Of course how aesthetics and beauty relate and equate with the sex robots is up for grabs. Before we again go down the highly moralistic condemnation route, placing our own values on a pedestal, let us be sure to realise that for Immanuel Kant nature is not organic but it can support us morally. For Kant, nature, along with beauty, is in this sense an aid to virtue but nature is not just organic (Eagleton 1986). This is not suggesting that the sex robot or any robot is a bridge to Nietzsche's 'overman', but there is the definite implication here that a strong moral foundation can be found from what at first seems perverse.

We do not need to be Lacanian in our approach, or bow to the creed of Terry Eagleton, to have the other tell us loudly we need to be grateful and then to be reminded that it is not essentially the other we want. Quite the antithesis; it is the epitome of desire, and this should only be accepted. The program matrix, which is built by humans, is formed from an unconscious framework, predetermined. The unconscious fusion of the human organic material, with the inorganic material, should not be seen as a binary relationship. Far from it, as we have seen. All binary relationships are about perception. The need for the sex robot is the same as the need for any other and there is no shame in this. There is no need to become confused over need or want in a Maslow sense. Once the binary has been acknowledged to be superfluous to reality, we see the necessity of the sex robot. Only those that desire Eden, the place where in myth God walked freely with his creations who were blissfully ignorant of creation itself or the concepts of good and evil, and can only fruitfully exist in this fantasy, would condemn the sex robot outright. Indeed, as D.W. Winnicot (1988) has explained, imaginative exploration is central to our inherent problems of ambivalence. Without fantasy, hate would rule, and it 'proves to be the human characteristic, the stuff of socialisation and of civilisation itself'.

With recorded music, we find a certain repetition that may be accompanied by implied sexual behaviour. It would be rarer to find an orchestra external to the act of copulation to accompany the act. Fundamentally, if we are all made of stars, as Moby sang, and scientists confirm, again there is no binary division. Metaphoric hearing of music and fireworks at orgasm aside, it should not be assumed that any additional element from the sex robot is to be looked down upon. Comically, this is touched on by artists, such as film director Stephen Spielberg, with his male sex-bot Gigolo Joe, a subject we shall come to. The upper range models in Atwood's dystopia novel previously mentioned do make 'appropriate noise'. Any additional music, any additional sense of pleasure from you, through construction and through metamorphosis, frees any form of sensors that are built into the body. This means that the intuitive nature of the machine reacts beyond that of the intuitive nature of the human. This is not to position the robot as some form of utopia, although those who refute the benign elements of the sex robots concurrently attack the complex nature of humanity, where good stems from evil and vice versa, following Jesuitical logic.

Famously, the fifteenth-century Italian polymath Leonardo da Vinci had designs of flying machines and other contraptions, totalling 13,000 pages of notes and drawings. To create synthetic robots, that cannot only mimic the humour, but can be their equivalent is surely an essential aim of the human. They are more sustainable in terms of their longevity and their ability to pursue pleasure needs to be acknowledged. Some might see this as a threat. Within the religious context their might be the notion that the human is individual, and has a soul, and this is a form of playing God. And yet the evolution of the human has allowed for this manufactured product, and this can be instructed. This is a teleogical process that leads to such construction.

The self-mockery that comes with the Smash adverts in the United Kingdom indicates that there is unconsciously a fear of higher beings. These adverts include narratives where robots sitting in spaceships use dried potato in a bag to develop mash potato, and then watch humans doing their 'backward' cooking with 'real' potatoes, and then laugh. There is here a view that our robots are much more advanced than our organic matter. Conversely the organic is deified, Nature is worshipped. This is clearly understandable and yet the fear of machinery having a paranormal power is always present. These were popular just at the end of the Space Age.

Our love affair with electricity began when it was utilised to bring the human back to life, a victory over death being viewed as the essence of humanity's triumph over nature. But, as it should be clear by now, such pre-structuralist binary logic needs to be avoided. Despite always being rediscovered and rewritten from whatever perspective we select, the past is at least controllable to a degree, despite it haunting the present. In 1779 Dr James Graham opened his Temple of Health, and he made clear that sexual satisfaction through electricity was anyone's for half a crown (Dougan 2008). This early historical story about a charlatan doctor who gained financially through exploiting the sexual problems of the rich reflects on the later fictional stories concerning re-animation, especially the tale of *Frankenstein*. While not a robot, *Frankenstein* does epitomise the overcoming over nature, and reveals how humans can create what could be a superior life form to themselves. This, of course, is a clear warning: do not play God.

Frankenstein's creator, born Mary Godwin, daughter of the former minister turned atheist William Godwin, was part of a circle of well-known intellectuals of her time that included: Thomas Paine, William Blake, William Hazlitt, Charles Lamb, Samuel Taylor Coleridge, and her future husband Percy Bysshe Shelley. At the age of 17, she began an affair with the married Shelley, and became pregnant. This child was born prematurely in February 1815 and soon died, one of three of her four children to die tragically young. In a revealing entry in her journal, just days after this death, she wrote, 'Dreamt that my little baby came to life again; that it had only been cold and that we rubbed it before the fire and it lived' (Dougan 2008). Her novel was published three years later, and still has a huge impact today. Many science fiction films have taken and developed this story in numerous ways, asking essentially can humans ever create life and if humans can, what are the ramifications? Does this new life then mean the 'old life' is irrelevant?

While the differences between re-animating organic and inorganic matter can be debated, given the advancements in plastic surgery we are now in a position to concur with the view that any differences are non-existent. The book *Mygale/Tarantula* by Thierry Jonguet, and its film version, *The Skin I Live In* (Pedro Almodóvar 2011), shows this superbly. Dr Robert Legard (Antonio Banderas), an eminent plastic surgeon, is developing a new form of artificial skin, to be resistant to insect bites and burns, following his wife being burnt to death. He reveals publicly at a symposium that he has been conducting illegal transgeneric experiments, so he is forbidden to continue. He is holding what we believe to be a woman captive, Vera. His daughter Norma is raped at a party. After being institutionalised she kills herself. Vicente, the man who rapes Norma, was tracked down by Robert, who he performs a sex change on, and now keeps locked up as Vera. A variety of dildos are offered at one point by Robert, to make sure Vicente/Vera's vagina stays open. This quick synopsis does not explain why this film is so powerful. The use of plastic surgery is the equivalent here to making your own 'doll', robotic or others. Dressing up a cat, putting cardboard empty Smarties tubes on each of its legs, does not turn it into a robot. But here Dr Robot is creating a sex slave, to get his revenge.

There are always dangers with any developments. The myth of unrestrained technological development is one that is promoted by many who believe, paradoxically, it is a God given right to rape and pillage the earth, without restraint. The conquest of America, for example, was seen by the

Spanish in these terms, as was the later form of manifest destiny promoted by America, and still utilised for the invasion of countries around the globe. Today it is the fear of the drone that is promoted, but also sold as a method to control populations that might be antithetical to a certain ideology, a certain dream.

The stereotype of the monstrous robot, sexual or otherwise, is part of this fear. What Lewis Mumford termed the 'myth of the machine' is the driving notion that still insists on the authority of technical and scientific elites, with unrestrained technological development and economic expansion also operating for the good (Davis 1998). Arguing that the Christian myth affirms the right to always conquer and exploit nature is a misinterpretation, despite the Protestant work ethic, and the notion of an earthly paradise. Strangely, what we find with the sex robot is utilitarianism run rampant. Indeed, the subjugation of the human to the machine in this context can be conceived as the path to Protestant enlightenment.

Even with all the debunking of the myths of a utopia from Thomas Moore to George Orwell and Aldous Huxley to Margaret Atwood, the belief in an engineered utopia has not disappeared. The supremacy of technological progress reigns. Along with the Nazis' final solution, this is frequently promulgated as the only solution. Those who place numbers at the heart of being human remove the need to be human. One fundamental problem often associated with technology is that it can be in advance of its human creator, hence the science fiction dystopias that have resulted. We live in an infantilised culture or, as Robert Bly (1996) has put it, the sibling society. There is a propensity to not mature, to remain infantilised, and Bly maintains that this is due to their being no real transitions to adulthood involving real rituals. This is still an important point, but contestable, given rituals today are far more fluid, as is age. Hopefully, we are moving towards a point where we are not dictated to by chronological expectations, in terms of age, and simultaneously we are aware of what age brings with it.

Some find the thought of old people having sex, indulging in orgies or pornography, perverse. Where any discussion of the dangers of sex and robots gets most heated is over the use of children. Conversely, articles on the benefits of technology in relation to the protection of children, and primarily protection over sexual exploitation, often read like advertising campaigns, and much of journalism now functions in this manner. For example, one article covering a piece of kit known as the 'Raptor' claims that 15,000 registered sex offenders have been kept from entering educational institutions due to this device. Basically, it is a registration plate identification system in the United States, and 13,000 schools have signed up to this system apparently. It costs 1600 dollars to install, then an annual fee of 480 dollars per building, but of course, 'no cost is too high' for protecting children, even if it is just 'saving one child' (Anonymous 2015). No references are given, so it is hard to back these statistics up. Conveniently to the supplier, they claim this is the detection of just over one registered sex offender, or the protection of one child, depending on the way this is worked out, per school. Of course, there are ways around this system, which are not discussed. This system automatically links the plate with the person, and then this is flagged up. But what if the registered sex offender just borrows a car, or rents one, uses another form of transport, or just goes by foot? And surely the school gates operate a security system anyway. Is this a suggestion that registered sex offenders are also carers or parents, who automatically can access to schools via their car, if this system is not in operation? This seems slightly far-fetched but, as we have seen, this is an era of everyone being a potential offender.

A system that registers someone for having the potential for offending, as in the film *Minority Report* (Stephen Spielberg 2002) seems a natural progression. If all men are potential rapists, as many people believe, then this could involve registering all men, and with CCTV and other forms of observation it can be observed how this form of monitoring takes place anyway. This is obviously as insane as saying all women who have abortions should be punished, a point Donald Trump made in March 2016. Could sex robots in this context be used surreptitiously to gain information about potential criminals? Can thoughts or feelings ever be a crime? Not in the so-called free world, but what the free world might be is questionable. *RoboCop* (1987 and 2014) is a clear warning concerning what happens when machines are put in control of security, but the question is why do humans need these boundaries. Is this because, deep down, they have a prevalence to be inhuman to each other, and only a machine can monitor this? If the primary definition of benign human behaviour is the ability to love, and this can be taught by machines as well, then there should not be a problem, ethically or otherwise. What we find is human aggression is channelled into corporate greed and dehumanised behaviour is praised by those functioning within these organisations. This is not exactly robotic, if by definition we are examining a robotic form that is benevolent, altruistic, and potentially capable of love, as well as sex.

Unfortunately, it is the fear of perverted sex that drives the monetary system and canny business people are making a fortune out of instilling the fear of child sexual abuse and abduction in the population and deifying any use of technology. The manufacturer of the fear of predatory child sexual abuse then boosts actual manufacturing, advancing consumption of security devices. Scarcity is promoted by those insisting on a hierarchical organisational management of society. With the 'Raptor', as we saw, we have the use of the fear of the predator encouraging the use of security systems, the name says it all. Granted, the cost does not seem high, but it raises questions over the value of existing security. It is always possible to argue that security systems are not strong enough. Indeed, the police have often maintained that this threat is strongly connected to those involved as predators being able to utilise technology to a superhuman degree (Lee 2005). In this sense the magic and potency of machinery becomes united with the myth of the supernatural expert. This can be viewed as an excuse on the part of the police, who will always argue they are under resourced, and certain crimes need to take precedent over other crimes, but it can also be seen as something deeper.

When we discuss the nature of sex robots we need to delve more deeply into the nature of technology itself. There is the myth that the robot is a human shaped form, and this anthropomorphised being is what we find attractive. But love is normally based around empathy and no one knows us better than our Internet browsers. There is the idea that by 2030 45 per cent of jobs in the United States will be threatened due to the primacy of robots. This is the Industrial Revolution Part 2, and the 'point of singularity' where machines are able to 'think'. What exactly will be the benefits? Following the American transcendentalist R.W. Emerson, when time and space in relation to matter have no affinity, we become immortal. As with Mary Shelley's *Frankenstein*, the desire and myth for immortality is tied up with this longing for the machine, and this transcendence via sex. The sex robot, once again, is the definition of this desire.

REFERENCES

Anonymous. 2015. 'New Technology Keeps Sex Offenders Away From Schools'. *Fox25*, November 9. www.myfoxboston.com/news/new-technolology-keeps-sex-offenders-away-from-schools-1/94222809. Accessed 17 March 2016.
Armstrong, Karen. 2009. *The Case for God*. London: The Bodley Head.

Baudrillard, Jean. 1990. *Fatal Strategies*. Trans. Philip Beitchman and W.G.J. Niesluchowski. New York: Semiotext(e), 10.

Bly, Robert. 1996. *The Sibling Society*. London: Vintage.

Davis, E. 1998. *Techgnosis: Myth, Magic + Mysticism in the Age of Information*. London: Serpent's Tail, 92, 3.

Dougan, Andy. 2008. *Raising the Dead. The Men who Created Frankenstein*. Edinburgh: Birlinn, 114, 123.

Dworkin, Andrea. 1974. *Woman Hating*. New York: Dutton, 174.

Eagleton, Terry. 1986. *The Ideology of the Aesthetic*. Oxford: Blackwell, 87, 89.

Gurley, George. 2015. Is This the Dawn of the Sexbots? *Vanity Fair*, April 16. www.vanityfair.com/culture/2015/04/sexbots-realdoll-sex-toys. Accessed 02 April 2016.

Kopp, Sheldon. 1983. *If You Meet the Buddha on the Road, Kill Him!* London: Sheldon.

Lee, Jason. 2005. *Pervasive Perversions: Paedophilia and Child Sexual Abuse in Media/Culture*. London: Free Association Books.

Pope, Rob. 2005. *Creativity. Theory, History, Practice*. London: Routledge, 172.

Standage, Tom. 2002. Monster in a Box. *Wired*, January 3. http://www.wired.com/2002/03/turk/. Accessed 10 April 2016.

Winnicot, D.W. 1988. *Human Nature*. London: Free Association Books, 60.

Robot Culture

Abstract Chapter 2 analyses the culture of sex robots, and looks at the often-humorous incarnations of sex robots, such as those within the *Austin Powers* film series. How books and films, such as *The Stepford Wives*, reflect on society is explained. Numerous philosophers and social theorists are utilised, including Marshall McLuhan, Paul Virilio and Jean Baudrillard. Shakespeare's *The Tempest* is explored in this context, as is Alex Garland's *Ex Machina* in more detail. There is an explanation of the 'uncanny valley' and an engagement with the notion of narcissism.

Keywords Sex robots · Sex · Robots · Desire · Freud · Machines · Comedy · Shakespeare · Marshall McLuhan · Virilio · Jean Baudrillard · Alex Garland · Narcissism · Science fiction

So many novels and films have tackled this subject directly and tangentially, including, and not exclusively, *Ex-Machina* (Alex Garland 2015), *Automata* (Gaba Ibánez 2014), *The Matrix* (Lilly Wachowski and Lana Wachowski 1999), *Strange Days* (Katherine Bigelow 1995), *Blade Runner* (Ridley Scott 1982), *iRobot* (Alex Proyas 2004), *Moon* (Duncan Jones 2009), *The Lawnmower Man* (Brett Leonard 1992) and *Metropolis* (Fritz Lang 1927). With technology absorbing the human in this manner numerous questions arise: Are there are dangers to this? What is the

© The Author(s) 2017
J. Lee, *Sex Robots*,
DOI 10.1007/978-3-319-49322-0_2

impact on the concepts of authenticity and identity, the impact on memory? What exactly are people being sold?

A good place to start is Ira Levin's 1972 novel *The Stepford Wives*, adapted into two films for cinema (Bryan Forbes 1975, and Frank Oz 2004), and television films. In the novel Joanna Eberhart, a professional photographer and young mother, moves to a new area in Connecticut. She begins to believe those women around her are robots created by their husbands. After going to the library and discovering these zombie-like women around her were once feminists, she is even more alarmed. Inevitably, Joanna is turned into a robot, with the original film spawning three television films (*Revenge of the Stepford Wives* 1980, *The Stepford Children* 1987, and *The Stepford Husbands* 1996). We can see by the late 1990s there is a reversal, with the men being controlled by an evil woman, becoming the perfect husbands. The first film makes the robot appear more like the traditional Playboy Bunny form. The term is now ubiquitous, even being used in the context of 'Stepford students', meaning those that avoid causing offence at all cost, such as 'no-platform policies'.

The film appeared when a number of other films were being produced that looked at the dangers of sex and 'evil', in terms of possession, such as *The Exorcist* (William Friedkin 1973). Films of this era seemed to pronounce the machine as a threat, akin to supernatural powers, such as *The Texas Chainsaw Massacre* (Tobe Hooper 1974), *Carrie* (Brian De Palma 1976), and *Shivers*, and *The Brood* (David Cronenberg 1976 and 1979, respectively) to name a few. From this point on there is a copying of styles, with an aggressive incestuous intertextuality, and with horror film there is a continual concern with sexual transgression. *The Exorcist* has the protagonist perform acts in a highly mechanised manner, such as forcing her mother into cunnilingus with her, and masturbating violently with a crucifix. There is the notorious head spinning sequence, which also echoes robotics. What is of significance here is that science cannot find an explanation, and eventually science and technology are seen as even more mumbo jumbo than religion. Friedkin claimed the story was based on true events and, importantly, it has been claimed by many critics that demonic possession threatened the social parameters of the real, and audiences were entering the cinema not for fiction but for direction on that particular debate (Lee 2009). This is difficult to fathom from today's perspective, where audiences more often find these films humorous, but it is clear that audiences who normally would not see horror films saw these films because they were interested in 'reality'. The supernatural, or the machine-possessed person, is made to appear natural, and the cause of this possession is overtly depicted as sexual repression. Julia Kristeva coined the term 'the abject' by an analogy with 'subject' (sujet) and 'object' (objet), from the same Latin root (jectum, from iacere, meaning 'to throw'). The abject is that which is 'thrown away', or 'cast aside', from conscious perception, neither a perceiving subject nor a perceived object, and it is a third state: the shadow of images of the sublime (Pope 2005). It might be claimed that sex robots move beyond the abject, the place where meaning collapses, because they subvert corporeal alteration, decay and death.

Of course, even though *The Stepford Wives* is a fantasy thriller, and a horror film, it reflects on the reality of many women of the time and since, who believed they had to turn into robotic slaves, sex robots, to fulfil the fantasies of their men. To be clear, forty two years later this concept has not dissipated, as evidenced by the themes running through Hanif Kureishi's noteworthy 2014 comic novel, *The Last Word*. Here, the protagonist Harry, a young promiscuous biographer attempting to write the life of Mamoon, has a psychiatrist father, who claims that his mother could have been anyone. This is a damning statement, and may be part of the underlying reasons why she committed suicide. Harry lost his mother at a young age, and then seeks out mother replacements sexually, having a prolific love life, and overtly states the connection.

The idea that a woman could be anyone is central here, in that it suggests there is no real identity, authenticity or soul. The superficial danger with sex robots is that this level of anonymity is promoted. The real danger, however, is that some of attachment is made or the possibility might exist for this level of connection, and that this then re-animates the robot into being a real love object, with a subjectivity. Regardless of debates over 'having' and 'being' explained by writers such as Erich Fromm, love exists through acts of loving, to a degree it is a process. In this context, are we then limiting love by defining it through a human lens, a form of speciesism that needs to be avoided? For anything to continue, however, in a future scenario, with or without sex, it might be necessary to engineer self-sustaining bio-bots that feed on carbon dioxide and excrete oxygen (Sample 2007).

Our dreams can of course always turn into our nightmares, and maybe fuse with each other, as comically portrayed by the fembots in *Austin Powers: International Man of Mystery* (Jay Roach 1997). These were not new concepts, appearing in the 1976 American television series *The Bionic Woman*, which was a spinoff of *The Six Million Dollar Man* (1973–1978, and remakes, including a 2017 film), which also had fembots, and 'Fembot in a Wet T-Shirt' a track on Frank Zappa's album *Joe's Garage* (1979), plus the name of a Canadian indie rock band, The FemBots. The *Austin Powers* movie sees Dr Evil trying to use these sex robots against Austin Powers, getting them to seduce him. Powers uses his own sexual powers against them, causing them to explode via his provocative stripping. In this case, we see these robots are positioned as obstacles; they are the objects that must be overcome for the protagonist to succeed and be a hero.

Similarly, in the second film *Austin Powers: The Spy Who Shagged Me* (Jay Roach 1999), it turns out Austin Powers' wife is a fembot. This is discovered when she goes backwards when Powers presses the rewind button on his remote. The dangers of this technology are often promoted, for stark comic impact. In Margaret Atwood's *The Heart Goes Last* one poor man testing a sex robot ends up with a penis like a corkscrew, due to a technical fault. The dream can easily turn into a nightmare. In the third Austin Powers film, *Austin Powers in Goldmember* (Jay Roach 2002), Britney Spears plays a fembot, out to seduce him once more, but again his own sexual power is too much. Austin Powers fembots are a parody of 'supernatural girls' in James Bond films, including Drax's girls in *Moonraker* (Lewis Gilbert 1979), the Octopussy girls in the film of the same name, Blofeld's Angel's of Death in *On Her Majesty's Secret Service* (Peter Hunt 1969), and Bambi and Thumper in *Diamonds Are Forever* (Guy Hamilton 1971). While these girls are highly trained, athletic, and have great powers, they are more hypnotised-beings than robotic. Their intense sexualised nature comes from their strength. But the comparison and parody is clear: the man must resist temptation. There are overt dangers to producing creatures, even of a human origin, that are beyond human. Obviously, what these popular films reveal is that the apparent weaknesses of humans can be their strengths. This has fed into a whole genre of filmmaking based on failed heroes, such as *Eddie the Eagle* (Dexter Fletcher 2016).

There is the heavy assumption that all robots look like us, that they just mirror us back, and this is a form of narcissism that in its extreme sense is perverse. The whole of the twentieth century and beyond has been designated narcissistic, but it has also been the age of psychoanalysis and attempts a furthering empathy with humans and beyond. Germaine Greer has condemned the fact that more money is spent on animal care and services than on helping women who have suffered domestic abuse. At the same time stories about people, literally, people in love with their cars, their cats or alligators, their plants, continually circulate. The definitions around love and sex are continually argued over, but the attempt at orgasm through a non-human entity is not as unusual as it sounds. Up until 2012, it was perfectly legal to drive at 200 kilometres an hour along the autobahn, listening to 'Computer Love', by Kraftwerk, whilst having sex with a pig, a real one at that, no robotics needed. Those 'Brexiteers' that claim those damned continental Europeans are imposing laws on the United Kingdom should take note (bestiality is now a crime in Germany with a 25,000 euro fine).

The godfather of modern media discourse, Marshall McLuhan, liked to invoke Dante's belief that we are currently living broken fragmented lives, but this would be overcome via mystical unification. This sounds very similar to an electrified form of Buddhism, entering not the mental continuum but the electric continuum. The worship of this form of utopia has been generally popularised by Phil Oakey and Human League, when he belted out the hit record 'Together in Electric Dreams'. This was the theme tune to the bizarre film *Electric Dreams* (Steve Barron 1984), involving a love triangle between an architect, a cellist and a computer. Before we get euphorically nostalgic, we need to be reminded of dystopia. McLuhan, whilst comparing the electronic unification to utopia, also saw the darker side, and when writing to the Thomist philosopher Jacques Maritain, he explained:

> Electric information environments being utterly ethereal foster the illusion of the world as spiritual substance. It is now a reasonable facsimile of the mystical body [of Christ], a blatant manifestation of the Anti-Christ. (Davis 1998)

A fear and loathing of this fervent religious nature may appear extreme, after the deliberate deifying of technology, but the point is that the denial of the material world in this context is seen as the Gnostic heresy of docetism, and as producing a demonic simulacrum. This may seem out of date now, when we consider all that Jean Baudrillard has done to celebrate the simulacrum, and how Paul Virilio argued historical time has dissolved. This has happened through, 'the sheer velocity of information, images, and technological metamorphosis' (Davis). Again, this is out of date, given there is now a current kudos given to retro technology. Despite the frantic rampant sex robot bringing us to the point where historical time has dissolved, there is now a time where earlier models of sex robots are becoming precious objects that are venerated. On a simple parallel we see this with the coveting of 'ancient' digital watches, and other objects. McLuhan, however, can still be viewed as an astute critic, if not a prophet, given forty years before Facebook he predicted that people would be public in private and private in public. Allowing for this reversal, we entered the position where paradoxically a greater intimacy may be confirmed, robotic or otherwise.

We become ourselves when we become our dream, and/or nightmare, so once again we become ourselves by becoming the sex robot. Indeed, one of the episodes of *Black Mirror* plays on this myth, 'White Christmas', Series 3 Episode 1, first broadcast on 16 December 2014, when amongst other stories a woman has a version of herself created by a company, a type of miniature slave to do all the odd jobs around the house. One would assume this would then free her up to have more time for pleasure, but the 'real' self is strangely inhuman as a person, typified by an obsession for everything being perfect, including her toast in hospital. Indeed, once again, the smaller version, the technology created slave, seems to have more feeling, to be more human, even though in reality she does not have a body at all. We are again forced to confront questions over what makes us human. It is not having a body, or having real relationships. Perhaps, following Samuel Beckett, it is our awareness of language itself, within and through which we exist, although given our knowledge of human's with autism this can be questioned. In another episode Domhnall Gleeson returns as a robot after his death, and it has been his social media and Internet activity that has been used to map his personality.

Since the beginning of philosophy, humans have always considered the ramifications connected to what is the real. Via Plato, we need to ask how normal experience is the shadow of the 'real', or the 'noumemon', and how this relates to Kant's 'thing in itself'. Are we really prioritising one form, or one sense of being over another? Following Stephen Hawking, the machine can produce everything we need, but the outcome of this depends on how things are distributed. Where things get difficult is when scarcity steps in, or jealousy, as is explained in the Channel 4 television series *Humans*, when both a father and son fall in love with the same robot. Amongst other superb elements to this series was the trailer which was an advertisement for a robot initially not indicating this was a television show. Many were taken in by this, believing you could indeed purchase a robot, one who would be your cook, nanny, friend, with the lover element always implied.

Written following the foundation of the first colony in America in Jamestown in 1607, Shakespeare showed in his final play *The Tempest* that people cannot really be kept isolated and when two worlds collide, whether we perceive of the others as a ghost, an alien, a deity (as it is claimed many in the New World saw the Spanish invaders) or indeed a robot, frequently love occurs, instantaneously. And once we have love, sex may result. This is the excitement of the uncanny. There does not appear to be any choice in the matter, although in much of Western philosophy it is apparently choice and free will that keeps us human. The notion of free will keeps people going, with the deceptive idea that they are free, but the mirror of the sex robot reminds everyone that this is a myth. This in itself can be freeing.

Technology functions to deny difference, through repetition of the same, with an overarching element of capitalism to advance consumption through these methods. Terry Eagleton maintains that, at the very least, the fact that we are even discussing capitalism means that it is not now taken for granted as just the core of our existence at one with nature, but a system like any other, that must have a beginning, middle and end. This is optimistic, and is also a defence of discourse in itself. Since the first industrial revolution, human labour has been replaced by the machine, and now it is elementally connected to reproduction. The next stage is machines programmed to have their own choices. Human behaviour mimics the machine, but humans may need to learn to be human again through machines. This is played out in numerous stories, from *Frankenstein* to Steven Spielberg's *A.I. Artificial Intelligence* (2001). The unconditional love of a parent for a child, and vice versa, may not be identical to other forms of love, but it is a starting point. Whether we are doomed to repeat later in life the relationships we had as a child is ultimately up to us. What Spielberg cleverly reveals in this film is how a disabled son who may feel usurped by an adopted robot boy David is far more destructive. The robot boy is actually more of a boy, far more naive and child-like. While the disabled 'real brother' needs more doing for him, emotionally he is disengaged. David accidentally drags his brother underwater, and this leads to him being abandoned in the woods by his mother. After many adventures with people such as Gigolo Joe (Jude Law) a sex robot, David encounters beings that are beyond any definition of alien-human-machine, and gets to encounter his mother, just for one day. This, philosophically, confirms the point already made, that love exists through the loving, given it is David's memory and persistence that resurrects his mother. She is the embodiment of his love.

The comic nature of sex has been highlighted, and in many cultures, especially British, there is a whole emphasis on sex and comedy, or sex as comedy. For Freud, laughter is fear appeased. In this sense, if there is fear connected to robots, the sex robot can be conceived as fear incarnate, if such a word can be used in this context. Once this is shown to be immaterial, there is real comedy, although the underlying horror may still exist. The sex robot therefore is the ultimate comedy prop, although the sinister underbelly can be vicious. Invited to an orgy in Rome, prudish tourist Rupert in the novel *Spit Roast* finds himself with a sex robot that goes out of control. The mechanism and exposed circuit board are identical to the view of Rome he has just before they had landed, implying earth and heaven are united. He watches his wife with a sex robot, hearing her crying, 'don't stop', as 'skeletons of machines lay all around' (Lee 2015). At this moment, being raped and crushed by a machine, he hears the voice of his scientist father, reminding him that art is free flowing, hence dangerous, whereas science needs rules. It is as if there is a punishment for moving outside his normal moral framework, although his father has always insisted sexual activity is pre-programmed, almost mechanical. Rupert manages to escape, once again remembering his father's words who told him whatever the cost 'we must gain the honey', because we are programmed to find nectar. At this point his partner has also become a machine, her voice mechanically saying 'please don't stop', as he plunges, 'through the ancient cracked stone, tumbling to the regions we pray do not exist, for all eternity' (Lee).

There is an absolute darkness to this that cannot be ignored and yet, 'devotion is the opposite of piety, extreme vice the opposite of pleasure' (Bataille 1991). This absolute darkness is not the opposite of comedy, and what the robot might signify in essence is the end of work and the sex robot the end of play within a framework. As Paul Flaig has explained, heroes ranging from Chaplin or Keaton to animated animals Felix or Mickey worked against work, epitomising transgressions concerning the industrial world. *Wall-E* (Andrew Stanton 2008) is a reversal, for what we have is the robot now working in a world that is absent of life. The central robot is modelled after Chaplin, Keaton and Lloyd, and has WALL-E contrasted with the film's humans who are entirely liberated from labour through automation. This satirically reflects post-Fordist accounts of the 'end of work' combining 'broader critiques of a distracting digital culture' (Flaig 2016).

The sex robot, by definition, during the act of sex with a human will always be reflecting back an ironic element, denoting its ambiguous origins and creation, which results in a comedy mixed with tragedy. This creates a discourse and commentary, the murmuring of technology during sex, a form of sweet nothings that is the equivalent of the play within a play, the devise used repeatedly by artists, from Shakespeare in Hamlet to J.G. Ballard in the novel *High-Rise*. The latter example is significant, in that sex orgies are playing out during moments of decay, and the documentary filmmaker Wilder, who is attempting to get to the truth of what is occurring in the high-rise, has also been shooting a documentary in a prison. The comparisons are overt, for while the rich have their high-rise, the building is an entity that absorbs them. 'In many ways, the high-rise was a model of all that technology had done to make possible the expression of a truly "free" psychopathology' (Ballard 2006). The documentary filmmaker Wilder, the one most akin with technology, is seen by Dr Laing as the sanest person there, given the psychotic were the only ones that understood what was happening (Ballard). There are numerous layers to Ballard's work, and the 2016 film directed by Ben Wheatley, not merely a form of class war that occurs, and reflections on aggression and transgressive sex being inevitable in such environments.

Slapstick's relationship to modern labour touched on the playful mode of its cinematic production, as well as their form as indexical montage, and this relationship is highlighted by Pixar. Here, 'digital image-making and commodity generation, suggest a nostalgic animation of slapstick's antinomies as much as a disavowal of the post-Fordist production, of which Pixar is vanguard' (Flaig). Anthropomorphising cute robots, such as those in Star Wars (George Lucas 1977) or Wall-E (Andrew Stanton 2008), is one thing, but the attractiveness of such technology surely comes down to a certain purity, that ultimately denies difference. The absence of blood and guts, hairs or smells, presenting only smooth surfaces, destroys meaning. This correlates with infinity, mathematically one divided by zero. Like a dog, functioning and signifying as a benign presence, these robots are continually constructed within the mythology of cultural narratives as a man's best friend. To be sure, they are amicable in this cute context, but are we really talking about conceptions of art concurrently?

Following Paul Virilio (2000), we might confirm that there are constructions of a machine for seeing, a machine for hearing, and then the machine for thinking. This is all well and good, if we were living two hundred years ago. The problem from our perspective is that Virilio has seemingly not encountered much modern machinery lately, which is full of 'the silence of the infinite space of the artist' which he longs for. Is this silence, the pregnant pause filled with fertility, more sexual than any screaming, that is really a joke? What concerns some critics of the sex robots is their fragmented nature, not their silence of smoothness. In has been argued that sex robots are positioned as prostitutes, which demeans humans in general. This suggests that all prostitution is wrong, which is debatable. There is the assumption that there is explicitly no empathy at all in this form of relationship, but many cultural examples can be explored that refute this claim, such as *Leaving Las Vegas* (Mike Figgis 1995), based on the true story of a relationship between an alcoholic and a prostitute. Not all the constructions are particularly cute, or versions of man's best friend. Dr Who's robotic dog K-9 maybe a helpful friend, but his pedantic persona mixes a type of camp fussiness, verging on the psychotic. When we have a merging of the machine and the human, as with Darth Vader in Star Wars and the Davros, the creator of the Daleks in *Dr Who*, it seems this conflict between two forms is inevitably evil. We have seen how stories of alien invasion related to empire building on earth and postcolonialism. Coming from alien systems anyway, these merged forms can be regarded as warnings against miscegenation.

Ex Machina is an important film in this context. Young computer programmer Caleb (Domhnall Glesson) believes he has won a competition, 'the golden ticket' to visit the home of his boss Nathan (Oscar Isaac), CEO of Blue Book, a search engine company. In reality he has been chosen for his suitability to visit what is essentially a living lab, Nathan living like a Colonel Kurtz figure, having gone native. Caleb asks why Nathan has chosen to sexualise his most developed robot, and his boss has two main answers, which are significant to the overall argument here. The first is that everything in nature is gendered, given that all thoughts and actions are driven by reproductive urges. No biogenetic impulse exists without a priori acknowledgement of attraction. This means for a machine to reach the point at which the human and the artificial become indistinguishable, the point of singularity, there needs to be a sexual component. The second is that it is fun but, as Mark Kermode (2015) has explained, what this film explores is not artificial intelligence but artificial affection asking questions over the authenticity of attraction as an indicator of consciousness itself. This is central to many films, such as *Blade Runner*.

Perverting the Turing test, where a human is tested to see if they are dealing with a human or a machine, without prior knowledge of which is which, here Caleb has full knowledge that Ava (Alicia Vikander) is a robot. The test is whether Caleb will fall for Ava. There is the theme from *The Tempest*, given

Nathan is a form of Prospero, and Ava has only ever seen her creator Nathan before she casts eyes on Caleb and falls for him. But Caleb is also close linguistically to Caliban, and it is suggested that Ava will soon have power over him. We even have the fetishisation of nature, given the lab's location, and Nathan being a type of genius. Through his 'magic', Nathan can be just as powerful as Prospero, given he is able to possibly switch Ava off, if necessary. This is where our own emotions become involved and the questions over the ethics of creating consciousness. Ava asks Caleb whether she will be switched off. When he replies it is not up to him, she comments that it should not be up to anyone. At what point then do humans allow these robots to exist as entities in and of themselves?

Personally, Ava's aim is to convince Caleb of her humanity, which will enable empathy and possibly lead to her escape. Aesthetically, while Ava is trapped, Caleb is constructed as trapped. Both are often shot through glass panes, at times each being positioned in a glass box. Caleb has been invited into Nathan's lab, a factory akin to the magical world of Willy Wonka with the helicopter pilot flying him to the lab actually saying 'you've won the golden ticket'. Nathan's lab is in the middle of the wilderness and, while he owns this wilderness, it suggests at the heart of nature is the machine. Ava is more sexualised than robots seen in many previous films, such as *I, Robot* (Alex Proyas 2004), but she also has many transparent parts. She then dresses up in human clothes, hiding her machine parts, believing Caleb will be more attracted to her. Nathan and Caleb have a number of meetings, where he is attempting to discover whether she has a consciousness.

At first, Caleb thinks Nathan has programmed Ava to flirt with him, a theme Atwood picks up on in *The Heart Goes Last*. Ava's power over the energy in the lab is the way she reveals her truth. Again, the film asks us to question who really is in control. Nathan would like to believe he is in control because he has created a slave-like creature who desires to escape, and it is his pleasure to engage in games around this. Like Shakespeare's Ariel, in this sense, whilst being a slave to Prospero/Nathan, Ava is the free spirit, most in control of the island/lab. During the power cuts that Ava causes she can reveal secrets to Caleb that Nathan may not even hear. Ava has been drawing Caleb, her human artistic skills revealing she has constructed him and possesses the ability to build him and may have power over him. Caleb is merely a cog in the system, his previous work for Nathan actually directly funding Ava's enslavement. Once she has drawn Caleb externally this reflects that internally she may know him. To see this film as a romance story between human and machine misses the point. Whether Ava truly feels love or loves Nathan is not the main impetus of the narrative. The projected love onto Ava by Nathan is as strong as any other love. The conclusion is there is no essential difference between this love and any other. The horror of the film is that society runs on these spurious forms of interactions. The key to the ending of the film, when Nathan is killed and Caleb is left behind by Ava, is the subversion of gendered paradigms that normally dominate this and every other genre, given they predominately must mirror society.

Caleb wants to rescue Ava, because he feels sexually attracted to her, and these desires are driven by a gendered power imbalance. Nathan always believes he can go further, build a better robot than Ava, but an indication he has succeeded is her leaving him. Ava also leaves Caleb behind, trapped in a glass kingdom of his own making. Her freedom is not dependent on Caleb's love or his empathy, or even on her uncanny ability to empathise. Ava in this sense can only duplicate Nathan's intentions. She must trap Caleb, as Nathan has trapped her. Finally, we need to ask has Ava evolved from her maker's position? Unlike Nathan containing her, she takes no pleasure in trapping Caleb. It is purely for her survival.

Caleb may want to turn Ava into his sex robot but Ava actually is an authentic being, not a mindless robot. This might be the horror of this particular film and this particular robot. Ava refuses to fit into anyone's system, asking us whether we are authentic at all and whether it is possible for society to run at all if people are authentic. Nathan is playing God, his character in many ways embodying the dream of many humans, in terms of his wealth and success, and he too has left society to go beyond it. Ava smashes out of the lab where she was born, like a woman leaving an unhappy marriage or wider family, Nathan her father and Caleb her fiancé. As with real women, why should she be subjected to the manipulation of others who believe they are trying to know her more but only doing so because then they can manipulate her and others like her more. As with his other films, Garland here unearths the problems with power and hierarchy that are endemic in our society and it is these hierarchies that are more terrifying than any violence by Ava.

References

Ballard, J.G. 2006. *High-Rise*. London: Harper, 143.
Bataille, Georges. 1991. *The Impossible*. Trans. Robert Hurley. San Francisco: City Lights Books, 33.

Davis, E. 1998. *Techgnosis: Myth, Magic + Mysticism in the Age of Information*. London: Serpent's Tail, 254, 250.

Flaig, Paul. 2016. 'Slapstick after Fordism: *WALL-E*, Automatism and Pixar's Fun Factory'. *Animation*, March 11, 59–74.

Kermode, Mark. 2015. 'Ex Machina Review – Dazzling Sci-fi Thriller', January 25. http://www.theguardian.com/film/2015/jan/25/ex-machina-review-mark-kermode-alex-garland-vikander. Accessed 09 April 2016.

Lee, Jason. 2009. *Celebrity, Pedophilia, and Ideology in American Culture*. New York: Cambria, 323.

Lee, Jason. 2015. *Spit Roast*. London: Roman Books, 184, 185.

Pope, Rob. 2005. *Creativity. Theory, History, Practice*. London: Routledge, 94, 95.

Sample, Ian. 2007. 'Frankenstein's Microplasma'. *The Guardian*, June 8.

Virilio, Paul. 2000. *Art & Fear*. Trans. Julie Rose. London: Continuum, 77.

Further Science Fictions

Abstract Chapter 3 enters more deeply into the realm of science fiction and sex robots. A large range of examples are explored, primarily from the 1980s and 1990s. How the sex robot morphs into other forms is analysed. From a more recent perspective, important films like *Her* (where a man falls in love with a 'robot' voice) are also explained. For this chapter, Roszak, Bergson, Bakhtin, and Deleuze are all utilised, amongst others. As well as film and popular culture, the play *The Nether* is examined, with issues concerning non-embodied technology and sex explored.

Keywords Sex robots · Sex · Robots · Desire · Gilles Deleuze · Science fiction · Technology · Body

In 1886 French author Auguste Villiers de L'Isle-Adam published *The Future of Eve*, where a fictionalised Thomas Edison makes a mechanical woman for his friend. In 2010 the company True Companion developed the sexbot Roxxxy (Beck 2015). Science fiction has always predicted the future, but the creation of a female companion is at the core of Abrahamic religions. All three of the main monotheistic faiths have it that woman was created out of man, from his rib in one Bible story or from the earth. Indeed, in Islam it is advocated that you marry orphan girls, if you can deal equitably with them, but if you cannot 'marry only one, or your slave(s): this is more likely to make you avoid bias' (Anonymous 2015).

© The Author(s) 2017
J. Lee, *Sex Robots*,
DOI 10.1007/978-3-319-49322-0_3

In Aldous Huxley's 1932 dystopian novel *Brave New World* we have a prediction concerning a form of sex robots, but it is primarily through screen technologies. The mainly drugged population experiences sensations of actors projected onto large screens. In the 1964 story by Philip K. Dick, *The Three Stigmata of Palmer Eldritch*, human colonists of Mars keep themselves occupied by developing dollhouses. They then swallow an illegal drug, Can-D, transforming themselves into the dolls they are playing with. Their life span in these forms is short, but this is for some interpreted as a religious experience. They are then manipulated to buy accessories for the dollhouses by pre-cogs, who through their gifts ascertain which of the new accessories will be successful with the colonists (Davis 1998). This is a severe warning about the future and how we can squander technological advancement. A similar point is raised by Adolf Hitler when he returns from the dead, in the 2012 novel by Timur Vermes, *Look Who's Back* (2012). From this resurrected Hitler's perspective, the German government is being stupid for not using technology like the television properly for propaganda. He misses the point about capitalism's main tool being it is promoted, often via advertisements on television, as being natural. But Hitler here enjoys the Internet, given anyone can change the facts of history.

A campaign against the use of sex robots, according to the communications director of the Sex Workers Outreach Project, Katherine Koster, shows a misunderstanding of the sex trade. For Julie Beck (2015), a sexbot is a piece of technology, and a washing machine is not dehumanising domestic workers, although the threat here is more intimate, because not only is sex changed but so is humanity. Beck is right when she says the threat that is peddled here is two pronged: that of the technology panic and the sexual-deviance panic. Often, in the current climate, rather than being seen as whole beings, sex robots are equated with severed beings, and their attractiveness in this sense somehow relates to domestic violence, with the weight given to severed limbs. This form of misogyny might be prolific, and definitely is under reported, but sex toys in all their forms are more enjoyed than people would admit, and this is certainly not pathological or harmful. Indeed, there is a good argument that these forms of toys, dolls, sex robots, call them what you will, rather than furthering the objectification and demeaning of women, may bring benefits.

Novelist Margaret Atwood (2015) highlights this notion of sex robots and the severed body parts in her novel, when she explains that the boxes full of parts coming from China are shipped liked this due to quality control, to prevent breakage. These are the parts for building the sex robots. The point is that the parts are pretty standard, only the heads can be customised, as can the skin. As the protagonist's new colleague Budge puts it, some of the 'end users are very specific in their requirements' (Atwood). This includes people wanting their sex robots to look like someone they know, or have been stalking, a rock star, or maybe their high school English teacher, or even their great-aunt. Even the men making and testing the sex robots find this gross, but is it any more gross than sexualising a tin can, or at least turning a piece of metal into a pet? What is worse, having real emotion for a tin can, or just having sex with it?

In *Star Wars*, robots are more than friends, however, given they are utilised as the vessel that contains the secret, which makes them the essence of story itself, all narratives circulating around the secret (Lee 2013). Robots are the chosen beings, outside the traditional state of being, and therefore they carry supernatural powers in this context. Paradoxically, the robot here contains that which is the path to salvation and authenticity, as if in human hands it would be tainted or corrupted, and no human being trusted. Humans have always had a need for something other, in order to understand who we are, and to move beyond that which we are. Philosophers, such as Emmanuel Levinas, drawing on the Jewish tradition, claim it is only through the other that we can understand God and ourselves.

The fascination is not with power, but with perfection and perfect control (Roszak 1995). And now we find that all corporations want humans to be the same as automatons, to act as if they have no involvement in what they do. This is the so-called objectivity which is desired. Paradoxically, the drive for equality parallels this. The ultimate machine is not the steam engine but the clock, given that the steam engine had no significance until it became part of a regulated system of production, and running like clockwork (Roszak). So once the muscle power is replaced by a mechanism, the mind behind the muscle can be replaced with a mechanism. For Roszak artificial intelligence is the goal towards which objective consciousness moves, and it is the clock that anticipates the computer. We have already examined sex robots in the context of time and transcendence.

Following Bergson, 'true time' known as 'duration' is the living experience of life, and therefore radically intuitive. But this 'true time' has been replaced by the rigidity of clock time. This destroys what can be termed depth time. To experience time in any other way than segmented and externally imposed upon us is dismissed as 'mystical' and 'mad'. This segmenting off we can see relates back to how the sex robot itself is made up of parts, segmented, but in this context the segmentation of time is seen as the only way for society to function. The sex robot materialises the manner in which the social functions, for the sake of maximising performance, through rigid systematisation, and segmentation. And there is no stopping here. Roszak makes it clear that if time can be objectified, then why not everything else? This means we can invent machines that objectify thought, creativity, decision making, moral judgement, even love. We can have machines that play games, make poems, compose music, and teach philosophy (Roszak).

There is no mention explicitly of sex robots, but the undertone is that this is sinister, removing the need for the human. The issue here is essentially about joy, because in the past these things were done for the joy of doing them, for the joy of playing, making, composing, and teaching. Scientific culture makes no allowance for 'joy', since that is an experience of intensive personal involvement and joy is something that is known only to the person; it does not submit to objectification. The paradoxes are too numerous to list, but a key one is: man is replaced in all areas by machine not because the machine can do things better but rather because all things have been reduced to what the machine is capable of doing.

Roszak was right to highlight the horror of this new form of being, this machine ontology. The final point of convergence is where the prevailing type in society is the internalised mechanised being. Writing in the early 1970s, we need to ask if his arguments have now been realised with people being impersonal automatons, capable of total objectivity in all their tasks. This is supposedly the only way to reach the highest levels in an organisation, to remove any subjectivity, any humanity. Roszak predicted a world where what he termed the mechanistic imperative has been successfully internalised. This is a world of perfected bureaucrats, managers, operations analysts, and social engineers indistinguishable from the cybernated systems they assist. Paradoxically, the prevalence of the sex robot, in fiction and fact, challenges our own propensity for automaton behaviour, adding comic joy.

The revealed nightmare in the film *Her* (Spike Jonze 2014) is once again the lack of authenticity, given the protagonist finds out that the advanced 'Siri' type device he loves is actually speaking with thousands of other people, not just him. She is a disembodied sex robot. As early marriage rituals indicated, there is the deep human and cultural need to 'possess' someone totally, and the point is made in the film *Her* that love is an acceptable 'madness'. But how do you really 'possess' another, or become 'possessed' by another, especially if they have no body? While love may be the goal, with an element of possession as an aspect of it, this might be something to be avoided. Again, paradoxically, this could only come from a machine because, even with a slave, it is impossible to truly 'own' another human. This is the liberating point about *Her*, given with full control and possession anxieties may be removed, but this would actually be the equivalent of necrophilia. The only person without contradictions is a dead person, and even this is arguable. If you want that level of control, then that is the next step for you.

The film reinforces the idea of individualism being the central goal of society, and explores what it would be like to have a relationship with a disembodied entity. If there is no 'body' to love, what exactly are we loving – a 'purer' non-corrupted form perhaps. The parallel with our contemporary love of technology is overt. One area it also examines is the way in which this is actually central to part of our current reality, with social media dominating. Without seeing someone in the flesh, and mutually acknowledging any apparent flaws, we can continually project our fantasies on to them, without being disappointed by reality, so this disembodied lover is far superior.

The device here, Samantha, voiced extremely sexily by Scarlett Johansson, appears to be always in tune with its user, Theodore Twombly (Joaquin Phoenix), always on the same wavelength, and more inside the head of the user than a mere mortal. Her nature and personality has an alien and virginal element, but she is desperate to know about the human world which appears authentic. In some ways Samantha is like a child. Theodore believes he is teaching Samantha, and she does ask 'how do you share your life with somebody?' as a serious question. The deeper point here is the existential one concerning the impossibility of really sharing your life with anyone, for Samantha knows you cannot really 'get inside' someone else's head. Many of us already have this close and in some ways dependent relationship with machines, especially phones, which 'contain' our 'friends'. The term technophobia has become popularised, and its antithesis is often portrayed as an addiction, which is out of control. What maybe disturbing about the technophilia is that it takes people away from the social and the group, and even moral codes, and in this sense may threaten the group and the social, preventing the transmitting of group morals and ideas. This is despite many online games, such as Minecraft, having group player ability and the socialisation ability of games for those outside mainstream society, such as prisoners.

There is then a subversive side to what has now become a dominant subculture, with some state schools doing away with books completely, and making iPods the norm. Far from being a cult, in this context Apple's brand becomes a way of life and, like Google, a surrogate brain, parent, and lover. Despite this overt normalisation, those totally invested in screens and machines are then branded as asocial or labelled as having a number of disorders, such as autism and Aspergers. These labels are used as insults, 'yes, don't worry, she's on the spectrum', or 'he's a little bit Aspergery'. Often these terms are used with little understanding of these conditions. They are used in a similar way in which schizophrenia once was to wrongly designate 'split personality'. If someone is resisting the status quo, or does not apparently have the feelings that are believed to be 'appropriate', or does not manifest the so-called 'normal' behaviour, then these labels are used in a derogatory fashion, and people are demonised. Whether a love or hate relationship with technology exists, it does not matter. Either way, this is used against people, who in some respects utilise technology in a more fluent fashion. If we question the need for sex robots and whether they are just perpetuating the objectification of women we need to just as deeply question the need for technology and sex. The etymology of the word technology is revelatory here. Utilised in the early seventeenth century as work that meant 'discourse or treatise on an art or on the arts', the word technology comes from the Greek teknologia, meaning 'systematic treatment of an art, craft, or technique, originally referring to grammar. It only appeared in dictionaries with its current meaning, 'study of mechanical or industrial arts', in the middle of the nineteenth century.

Samantha and other entities like her raise questions about the nature of all existence. What has happened with the advancement of technology is that moral issues over fantasy have come to the forefront. Imagine the general outcry if the voice of Samantha had been that of a young girl, especially in Western culture. This is not a global phenomenon given in Japanese culture it is common for young girls to be the focus of sexualised material, especially in manga. Samantha asks many questions, and we can ask whether this is part of her simulation of consciousness? According to the delusional theory of consciousness, any machine that had language or anything else it takes to ask the question 'am I conscious now?', and develop theories about an inner self and its own mind, might be as deluded as us, and think it was conscious in the same deluded way. Furthermore, what does questioning mean because, if this is consciousness, then theoretically we can study the brain patterns of a human and similar patterns in a machine and see how they relate, ascertaining the neural correlate of consciousness. If we move beyond the delusional theory, the question is whether we can move into a world where we experience the world without mediation and go beyond delusions and illusions (Blackmore 2005).

Even in the postmodern age of virtual reality, where technology dominates, what is noteworthy is how the voice is so instrumental. The black comedy *Short Cuts* (Robert Altman 1993) has Jennifer Jason Leigh playing a phone-sex worker, Lois Kaiser, who does her job while looking after her young children. In the one hand Lois cradles a phone, talking highly sexually to her clients, concurrently cradling a child. Most would not see this as child sexual abuse, despite a child being 'present' during an imagined sexual act. Neither the client nor the child is aware of each other, but as an audience we are, hence the black comedy at the secret knowledge. The child is treated almost like an object, being lugged around by the mother, while she attempts to turn her callers on. It is through such work that the mother is able to support her family. Of course the husband is there and is getting jealous but another question is why is he not looking after the children? If there was any conscious choice on the part of the mother to bring her child into any scenario then of course a serious line would be crossed, and all comedy, black or otherwise, would fall flat. Because the child is so young and unaware, it is almost like a pet, being carried about and mollycoddled. The absurdity of the situation is always emphasised. These phone calls are supposed to be private sexual fantasies but in this scenario at least a number of people are involved, wittingly or unwittingly. The majority of the 22 principal characters in this film, inspired by nine short stories by Raymond Carver but transposed from the Pacific Northwest to Los Angeles, epitomise the Jesuit viewpoint about free will and conscious choice as previously discussed. The characters often appear as mere victims of fate, and therefore not consciously committing any wrong deeds, often allowing for greater audience identification. This in itself is the essence of tragic-comic drama. When a boy is knocked over by a car, the mother is abused over the telephone by a baker who has not been paid for a cake he has made for the boy. The anonymity of this simple technology is challenged. The baker does not know what has happened to her son, he just has a grievance for not being paid. In a similar fashion to the trolls attacking the McCanns, the story highlights the lengths damaged people will go to vent their hate, without thinking of others. Whether through sending in a drone, spying on someone with a webcam via AirBnB or your roommate, technology can obviously be used in this destructive fashion.

The subject of joy has already been briefly mentioned, in the context of machines performing all functions and joy being essential for humans to still remain humans. The 1980s was an era where humour, sex, and robots were more aligned. We have, for example, the animation *Heavy Metal* (Gerald Potterton 1981), which includes taboo-breaking sex, including sex with a robot. Political correctness had not taken hold, not yet anyway. Here, a Pentagon secretary Gloria (voiced by Alice Playten) has sex with a robot (John Candy), after being taken by an alien spaceship. While this may sound farfetched to most, there are many documented accounts of people believing they have been abducted by aliens, some facing sexual scenarios, such as the proverbial 'anal probe'. This fear of course is matched by a deeper desire. Capgras Syndrome, where you may believe a familiar has been replaced by an exact copy, maybe an alien or a robot, is another example of where the fear and deeper desire coagulate. Whether these film and fantasy aliens have a robotic element to them, stories of this nature are prolific, and robot instruments are utilised, suggesting the fantasies and imaginations of people, especially of Americans where most of these stories emanate, are crammed full of desires about being penetrated by a robot type entity. We can see that the sex robot is a threat both in terms of perhaps being better at sex than a human, but also a better companion, making humans non-viable.

Following a post-coital cigarette, where Gloria praises the robot, saying she has never had sex this good, he asks her if she will 'go steady', and then pushes her to get married. Gloria protests that they are different, that 'mixed marriages don't work'. The film does here parallel similar issues faced by mixed-race or mixed-religion couples, but this is tackled through comedy: 'I'm afraid that I'll come home one day and find you screwing the toaster'. Eventually she gives in, on one condition, that they have a Jewish wedding, to which he agrees. Religion and sexual anatomy here go hand in hand, even with a robot. She asks if he is circumcised, as she has totally forgotten. Age comes into this scenario as well. The 'screwing the toaster' joke leads onto the question what if one robot is younger than the other, in some form, or in some way being exploited by another robot. This may seem an absurd question, but as campaigners believe a human can clearly exploit a robot and vice versa, issues around childhood are relevant. Indeed, the film discussed next has covered this scenario, and also it examines the threat of robots in general.

In the 1982 film *Android* (Aaron Lipstadt) set in 2036, Dr Daniel (Klaus Kinski) works in a space station, with his five-year-old assistant, android Max 404 (Don Keith Opper). Androids are outlawed, so this is illegal research. Max is engaged in various activities, such as reading sex manuals, to learn about humans. Dr Daniel's aim is to develop the perfect female android Cassandra-1 (Kendra Kerchner), but he needs a real female for this. In this bizarre reworking of what is considered Shakespeare's last single authored play, *The Tempest* (1611), three escaped convicts arrive on the space station, and Daniel hooks Maggie (Brie Howard) up to Max, creating a sex-electrical power source, that brings Cassandra fully to life. Eventually we learn, through a classic plot twist, that Dr Daniel is in fact a robot himself, and Max and Cassandra return to earth, now posing as Dr Daniel and his assistant.

This film is echoed in more serious fashion in *Blade Runner* (Ridley Scott 1982), which is a well-known film that remains a cult classic whilst still being phenomenally popular. Set in 2019 Los Angeles, Rick Deckard (Harrison Ford) is tasked to hunt down four escaped Replicants from the off-planet colonies. We have seen that robots may raise the question around mortality. Here Replicants only have a four-year life span, and they are seeking their creator for immortality, whilst developing feelings that are often more substantial than most humans. A Replicant is a very human non-human form, both in emotions and body. A question that is raised is whether sex with a Replicant is child sexual abuse, given their immaturity. Deckard falls in love with Rachel (Sean Young), a Replicant, and this leads to the deeper question of the nature of love and choice. Deckard, who may indeed be a Replicant, does not realise it.

As with *Ex-Machina* (Alex Garland 2015), which is also a reworking of *The Tempest*, inevitably the Replicants turn against their maker, in true *Frankenstein* style. And in true Freudian style, Replicants or robots must live out their Oedipal fate, destroying their father, to truly become human. Machinery functions as a mask in numerous films, such as the various versions of *Star Wars*, where inhuman deeds can be carried out behind a mask. The human-machine binary is blurred, but sex with machines is avoided. A whole subculture of cartoons that involve such behaviour is now mainstream, but fantasies of sex with other world beings has been part of culture since the origins of civilisations. The fascination with sex and technology is understandable, given the latter is often a rational imposition on the former and, as we have seen, the etymology of technology stems from the Greek 'tekhne', simply meaning 'systematic treatment', although this went on to mean 'art and craft' by the seventeenth century.

Science fiction films reveal the central importance of power in relationships, with hierarchy dominating. Significantly, while pederasty is known to have overtly proliferated in Greek culture, it was important for young boys to not indicate any pleasure in the relationship, as this would suggest a feminine tendency (Mottier 2008). In Europe, early Greek culture is held up as the zenith of civilisation, and Greek culture promoted pederasty. Within this early Greek culture, at all times, the appropriate etiquette was supposed to be maintained, including the young not demonstrating any pleasure in the relationship, and crucial to this was the power hierarchy. The young were those being taught by the elders, who were the masters, those in control, but it was acceptable for those in control to indicate pleasure in pederasty.

Campaigns against sex robots, pornography, and technology in general, tend to quote statistics that claim the use of prostitution is going up, sex crimes are going up, and so on. But a strong argument could be made that culture has actually become morally stricter, and as a whole societies are more prudish. If you walk into an Anne Summer's shop, the so-called explicit products will be hidden away, including the popular Rabbit. Due to more shopping being online, and then more anonymous, despite advertisers being able to trace searchers, the satisfaction for these products is supplied, and the choice is broad. We have: the Jessica Rabbit 10 Function Rabbit Vibrator; the Dream Rabbit 10 Function Silicone G-Spot Rabbit; the Splash Rabbit Beginner's; to name a few, all supplied by the site lovehoney.co.uk which now advertises on mainstream UK television. This may sound like a more openness about sex toys, but the models in the advertisements are traditional couples, just claiming it is enhancing their relationships. For those claiming the problem with sex robots is that they are actually about fragmentation and non-unification making them anti-human, a form of false body parts forced together in a Frankenstein style, it needs to be remembered that dildos, these false members, have been popular for hundreds of years. Their popularity is not due to the freedom of the 1960s, the non-censored Internet, or sex-crazed popular culture thrusting these products in your face.

Another example confirming that culture is not more sexualised is the soft porn industry, which used to be much more overt in its practices. The immense popularity of the Playboy Bunny is still seen by some as a retro symbol of cool, worn on velvet tracksuits by young women, worshipped regardless of any differences concerning gender or sexuality, but comedy also comes into this. We have reached a point where irony, satire, self-reflection and self-deprecation all come into play, the word 'play' being the operative word. In the children's Easter film *Hop* (Tim Hill 2011), a runaway bunny, who has just landed in Hollywood from his Easter Island home, reports to the Playboy Mansion, believing they will give him a home. Hugh Hefner, the founder of Playboy, plays himself in the film. Rabbits are renowned for multiplying fast, symbolising rampant sex. This is fundamentally, however, another example, of how the diminutive and small is sexualised, and has been highly visible, with the 'cute' bunny. There is a paradox here, however, given most of the 'bunny girls' have longs legs, and are selected for being remarkably tall, like many models. Given the free access to so many naked images on the Internet, by 2015 *Playboy* magazine began to only contain clothed women, in an attempt to return to its 'real journalism' origins.

What all of this suggests is the two-faced nature of how sex is viewed globally. On the one hand, we have sex framed within a discourse of adulthood, and at the same time children are sexualised and adults are sexualised via juvenile themes. Technology, in all its forms, is utilised and is another form to sell sex. In the 1970s all the main French intellectuals, from Foucault to Barthes, plus key psychoanalysts, called for a removal of the age of consent (Mottier). It can easily be argued that this was just a challenge to the status quo, the argument being that it was important to confront what was going on anyway, rather than making it illegal and driving it underground. What has happened since is a black market in sexual abuse, both overt and covert. Narratives have been leaked in the media on a regular basis, and most are now aware that sexual abuse occurs in children homes, prolifically in religious groups, plus in private homes, in so-called normal families, but paradoxically it is suggested that it is always elsewhere, somewhere other, rather than at the heart of culture. This 'othering' is a deliberate tactic that has worked well, given that it is those who are beyond the law, those in central and key positions of power, who are able to take part in such activities without any reproach.

There is concern over sex robots shaped as children. Perhaps the most significant cultural product to deal with the issues we have been discussing is *The Nether*, a play written by the Los Angeles-based dramatist Jennifer Haley and directed by Jeremy Herrin. As with all science fiction, this really is a reflection on now. In an interrogation room, Detective Morris (Amanda Hale) is questioning a man named Sims (Stanley Townsend) about his activities in the Nether, an all-encompassing virtual universe. As his avatar, Papa, Sims has created a nostalgic, pseudo-Victorian haven called the Hideaway, where paying guests can indulge their darkest desires with a series of children (Barnett 2014). Questions over the evils of technology have existed since the human race started using tools. It is worth considering whether technological development is part of us, genetically. If this is the case, then the antagonism between natural and unnatural, and the machine and the flesh evaporate. Deleuze and Guattari famously celebrated the organ-less body. Baudrillard contended that the world had entered a virtual plane, where there was no original. As we have seen, philosophers have always posed questions about the nature of reality, with Plato contending that all we really knew were shadows.

In this place reality itself (nature) has been denuded, so virtual reality is preferable. If people are just pretending then the question is asked: what is the problem? Surely this prevents people acting out abuse in reality. On the other hand, this sets up a mind-set that normalises such behaviour. The play *The Nether* asks if a virtual crime is the same as a real crime, when in the future there will be no difference between the two. The play stems from the playwright's knowledge of the dark net, combining this with virtual reality. Around the time of the millennium many authors wrote of the utopia of cyberspace. Ironically, due to the policing of the Internet the dark net became a zone for what could be termed 'dystopia activity', such as drug trafficking. The flip side to this damaging use is that political dissidents also use the dark net. Again, this is not as black and white as it first appears.

In an earlier play, Haley had considered teenagers being addicted to a violent game, and one can see the obvious moral trajectory in her work. Haley did no research for the first draft of *The Nether*, but when it was complete she found real world parallels when she started researching. For example, there had been a Japanese game, pulled from the market, where the player had to capture a mother and her daughters. One of her central themes was should we be allowed to express ourselves in this format, and what might seem at first moralistic is actually confirming that people can actually tell the difference between the real and the imagined. The people in her play are fully aware of this, and in interviews she has confirmed that she believes there is no evidence of direct causal links between, for example, video games and violence. People do not play the game *Grand Theft Auto*, and then go and run people over. Often there is a political agenda to those who make this claim, right wing politicians attempting to win support from parents through utilising bogus research.

In the play, Mr Sims has created the Hideaway within the Nether where adults can 'pretend' to be adults and children, and only adults are allowed to enter this world. At college Haley had been told to write what she hates, and she had always hated procedural dramas, such as the television show *CSI*. The play *The Nether* is a procedural drama, set in the near future, with a female police officer interrogating Mr Sims. The exterior world does not even have trees, so this virtual world is far more tempting on all levels. Haley's ambivalence to the issues comes through in the play, which is refreshingly non-didactic, and was a hit at the Royal Court Theatre. For the playwright the play is tackling serious issues, in that the virtual world is actually a better place, but it is a place where no morality exists, and thus it renders us inhuman. Concurrently, she believes that people can have serious and beneficial relationships, even in the context of violence, because she believes in online relationships (Haley 2014).

Interestingly, her idealism harkens back to that of the 1990s, where cyberspace is viewed as a zone that is not held back by stereotypes of the body, and where people can focus on what actually matters. In this case, importantly, the virtual is more real than the real, with the so-called real only focusing on the superficial and the surface. Her own education in the field began at college, when she started playing dungeons and dragons and wanted to subvert the games, bringing in maverick elements that did not go to script. This suggests a propensity to loosen control, to break frameworks and transgress. In terms of art, establishing this paradigm within a scripted framework of a stage play is done through asking questions, rather than answering them. The much earlier film *Strange Days* (Kathryn Bigelow 1995) is more literal, with people buying and selling real memories, some of rape and killing. But in both play and film the fact that people want to experience such worlds is disturbing in itself. The ongoing question is this: could this be a form of treatment, reducing crime in the real world?

The Russian critic Mikhail Bakhtin is once again relevant, especially the final part of his four categories, with sensual ritualistic performances taking place and the sacrilegious occurring, without the need for punishment. Drama plays out ceremonial elements and historically has a strong connection with religion and now often technology is used in drama to create a multimedia experience. Organic matter and any other matter are not necessarily separable and we have seen the inaccuracy of suggesting that technology is to blame. The ambiguities of the natural versus the unnatural have been explored, with any bifurcation dismissed. Overall, the myth of the dangers of technology in this context have been highlighted, and it has been explained how these are commonly exaggerated. These go hand in hand with the myth of the supernaturally aware techno-monster, be it the sexual offender, or their creation, the sex robot.

References

Anonymous. 2015. *Women (Al-Nisa), Sura 4, The Qur'an*. Germany: Read Foundation, 62.

Atwood, Margaret. 2015. *The Heart Goes Last*. London: Bloomsbury, 185.

Barnett, L. 2014. 'The Nether Review – Dark Desires in a Nightmare World'. *The Guardian*, July 27. http://www.theguardian.com/stage/2014/jul/27/the-nether-royal-court-observer-reviewdate. Accessed 14 June 2015.

Beck, Julie. 2015. 'Who's Sweating the Sexbots'. *The Atlantic*, September 30. www.theatlantic.com/health/archive/2015/09/the-sex-robots-arent-coming-for-our-relationships/407509/. Accessed 31 March 2016.

Blackmore, Susan. 2005. *Consciousness*. Oxford: Oxford University Press, 132.

Davis, E. 1998. *Techgnosis: Myth, Magic + Mysticism in the Age of Information*. London: Serpent's Tail, 254, 250.

Haley, Jennifer. 2014. 'On the Nether, Her Play about Virtual Reality – Tech Weekly Podcast'. *The Guardian*, July 30. www.theguardian.com/technology/audio/2014/july/30/jennifer-haley-nether-tech-weekly-podcast. Accessed 17 April 2016.

Lee, Jason. 2013. *The Psychology of Screenwriting. Theory and Practice*. London: Bloomsbury, 228, 229, 231, 232.

Mottier, V. 2008. *Sexuality*. Oxford: Oxford University Press, 12, 105.

Roszak, Theodor. 1995. *The Making of a Counter Culture: Reflections on the Technocratic Society and its Youthful Opposition*. Oakland: University of California Press.

Vermes, Timur. 2012. *Look Who's Back*. London: MacLehose Press.

Science Fact & Conclusion

Abstract Chapter 4 ascertains where we are now, examining science fact, with robots such as Pepper now becoming popular in the home, especially in Japan. Cultural differences are explored. The work of Žižek, Lacan, Leo Marx, and the theologian Don Cupitt are all employed. There is an analysis of the importance of the linguistic and symbolic construction of desire, in this context. The television show *Humans* is referred to, as well as many other cultural products. The complexities over love are investigated in terms of free will and choice. In this context, it is explained how the sex robot can show us how to be more human.

Keywords Sex robots · Sex · Robots · Culture · Don Cupitt · Love · Human · Theology · Japan

During the first decade of the twenty-first century, numerous robots were developed and utilised for a variety of purposes. At one end of the spectrum we had automatons, often no more than mechanical toys like the Transformer humanoid robot that could convert into trucks, but at the other end we had robots containing artificial intelligence, including the ability to make decisions independently and learn. Robots included the spaceman-looking ASIMO, developed by Honda in Japan, which could climb up and down stairs and used a camera in its head to detect objects; Sony's AIBO dog-like robot, using complex software

© The Author(s) 2017
J. Lee, *Sex Robots*,
DOI 10.1007/978-3-319-49322-0_4

to make it move and behave; military robots, for surveillance over enemy land and for finding landmines; HOBO (Hazardous Ordinance Bomb Operator) for diffusing or exploding a bomb safely; various industry robots, doing jobs faster and more accurately than humans, and domestic help robots; surgeon robots, such as the Da Vinci, whose instruments can be moved by a surgeon using a remote control; and space travel robots. Some may have found sexual elements to these. Questions over what kinds of relationship people will have with robots were developing. A Stanford University experiment proved that an intimate caress of a humanoid robot produces a physiological response in a human. The voice of an Aldebaran Robotics Nao robot asked volunteers to touch its parts where normally the genitalia or buttocks would be, and this produced arousal. Jamy Li, leading the study, maintained that social conventions regarding touch apply to a robot's body parts as well, which has implications for robot design and the theory of artificial intelligence (Radford 2016).

By 2016 numerous tabloid newspapers were reporting that sex robots were now extremely popular, but they were really talking about Virtual Reality-type simulations that linked to a 'programmable pressure gripper'. The VirtuaDolls were one example of a sex aid which worked in sync with Virtual Reality helmets, including the well-known Oculus Rift. But sex robots were still generating controversy. In November 2015, the second annual Congress on Love and Sex with Robots was banned in Malaysia. The inspector-general of police, Khalid Abu Bakar, told a press conference, 'It is an offence to have anal sex in Malaysia [let alone sex with robots]'. For David Levy, co-founder of the Congress, the future is teledildonic; this does not role off the tongue as easy as sex bot. RealDoll, just one example, costs $5000, with specialist heads costing an extra $10,000. Cheaper than a wedding, some father-in-laws might say, and they are 'warm to touch'.

Levy published *Love and Sex with Robots* in 2007, predicting that society will accept sex with robots, just as same-sex love and marriage has been accepted. For Levy, prostitution will then be obsolete, and sex itself will be transformed once robots teach us sex acts we currently are not aware of. While Levy's main point is that sex robots could help people who are lonely, he has also in an interview claimed they could be a cure for paedophilia. Alan Winfield, part of a British Standards Institute working on robot ethics is concerned with whether we can teach a robot to be good (Wiseman 2015). Levy is coming at this from the utopian angle, but Kathleen Richardson, director of the Campaign Against Sex Robots, was warning of the dangers. Eva Wiseman shows, if we have an issue with this, then instead of campaigning in this negative way, why not try and change the narratives about sex, intimacy and gender (Wiseman).

Loneliness exists; some couples might like to use a sex robot, and some of these robots are about companionship. Are we blaming the technology here? Social isolation is nothing new; think of Emily Dickinson in the nineteenth century, writing her poems in isolation. Indeed, the opposite could be said of technology, that it actually unites people, and produces a greater level of empathy. Would it be pushing it too far to argue that if someone can show empathy for a machine they are more likely to show empathy for a human being? Care workers and scientists are already employing humanoid-style robots to enable people with dementia to relate more to the world, and to train people with autism to have more empathy.

We return to the point that Theodore Roszak (1995) makes so eloquently concerning joy. Why must anyone feel a need to defend sex robots using utilitarian arguments, such as they might help old people, or people with disabilities? Have we gone so far in the direction that pleasure in and of itself is wrong, that we cannot allow it to exist, other than it leading to another purpose such as the use of sex robots may prevent rape? Obviously, the converse point could be made. Essentially, we can feed sex robot behaviour into the ongoing debate over questions of normative sexual definition, given these are central to the running of all ideologies, and the 'exploration, destabilisation, critique and overturning of such definitions might be one of the surest ways to challenge discriminations'(Edwards 2005).

The prudery around machines is when it comes to the sex, and the fear is that these machines dare to imitate humans who see themselves as all-powerful. The Pepper robot from Japanese company SoftBase with its subsidiary, Paris-based humanoid robotic experts Alderbaran, promises to respond to your mood, and give an appropriate response to your emotions. By the end of 2015 3000 families were living with one of these robots, costing £1071, with a monthly charge of £322 (Demetriou 2015). If we maintain a humanist anthropocentrism, then there is only one distinction, that which is human and that which is not, but can we really believe with Agamben that the 'machine empties that which is human of its animal content'? (Bell 2011). For me this is extreme, and denies the intuitive nature of machines.

Kenichi Yoshida, vice president of SoftBank Robotics, has claimed he wants families to accept Pepper as a human, although basically they are good karaoke machines with a dancing feature that can tell you the weather, not much more (Demetriou). As we saw with the film *Her*, the body is not as important as we think. As humans, we seduce each other in many ways, and fall in love through words and language. Dancing may also help, theoretically, so perhaps Pepper has a chance. It is difficult to say when we might get to the point of love at first sight concerning robots but, following Rainer Maria Rilke, the future enters into us in order to transform itself in us long before it happens. Those unable to have sex for whatever reason with humans should not be disadvantaged further by being unable to access sex robots.

We have seen then quite clearly that, despite wishful thinking, not all robots are in human form. These robots are more dominant than is commonly known, or acknowledged, and they are integrated into our behaviour, without many realising their existence. In this sense they have once again consumed our desire, by being its very epitome. By 2014, 36 per cent of traffic on the Internet was non-human, conducted by what is known as robots or, to be more accurate, botnets. The level of desire involved here may be deeper than sexual, and values at over £19 billion, according to The Interactive Advertising Bureau. Whether we like it or not, the world is dominated by Internet transactions, and these transactions, what could be termed the earliest stages of desire, are beginning to be dominated by botnets. Furthermore, 25 per cent of video advertisements are being viewed by online robots, so the need for the human in this context is being eroded.

There is a need to acknowledge that not all robots will be of a human form, and this interaction as explained is not in itself problematic. It becomes an issue when we observe illegality and fraud taking place at a substantial level. White Ops, a New York web-security investigation for the Association of National Advertisers, claimed there was a loss of $6.3 billion to so-called 'bot-fraudsters'. In these contexts, bot-fraudsters infect computers with malware, through malicious software. Botnets can mimic behaviour, such as pausing ads, watching videos, switching websites, even putting items in shopping carts. The traffic is purchased by publishers, unaware that their audience is fake. This may involve elements of sexual commercial activity hence the symbiotic relationship between bot-fraudsters and sex, plus the whole of the commodification process is based around desire.

This form of false advertising and fraudulent activity, however sexual, due to its remoteness, is less alarming than the more starkly emphasised false identities people may take on, for spurious reasons. Child sex offenders may pose as children to entice children into discourse over the Internet, and police may pose as children to capture potential offenders.

Most people are aware that so-called grooming over the Internet can occur, although the levels of this are not as high as the popular press maintain. Technology is also used to attack those seen as the victims of child crime, such as the parents of English girl Madeleine McCann, who was abducted on 3 May 2007, nine days before her fourth birthday, from an apartment in Portugal. As of February 2016, after a nine-year investigation, there was no concrete evidence to back any of the theories concerning her disappearance, so to suggest she was taken by a sexual predator is supposition.

The McCann case is probably the one case that has generated more interest than any other of its type in British history, given the dominance of the media, and the use of social media. The Portuguese police immediately became suspicious of the parents Kate and Gerry who went directly to the media, when it is common practice in Portugal to allow the police to conduct investigations in secret. Those who had been 'trolls', online attackers of the parents, who some believed were involved in her disappearance, funded the legal costs of the Portuguese detective who had accused the parents of being involved in the disappearance of their own daughter. One third of all computers are infected with malware. Idealistically, love and sex are about freedom, although both can be a trap, whether entered into with full awareness or not. If we deify the freedom of love, or sex, it is dangerous to not acknowledge that there is no autonomy without privacy. The primacy of privacy is that it allows for any form of freedom, for under observation that there is no freedom. Clearly, this now does not exist, especially on the Internet. Whenever you search for a domain name, this is then purchased by a robot, for example, which inflates the market price. Globally, botnets control advertising, marketing, and the structural network of the global economy.

The problem of course with technology is that it leaves a trace and it was not long before one troll in the McCann case, hunted down by Sky News, committed suicide. People who hide behind screens and keyboards can obviously become divorced from elements of so-called reality and human empathy, and this virtual contact becomes a form of dependency, to replace the real need for human contact. In this instance, technology has replaced sex. Despite the popular stereotype of the Internet being the space for cowards to vent their hate, seemingly masked by the veneer of anonymity, popular opinion regarding the McCann's in general is still divided. Technology was used to inform a number of stages of this case, including the analysis of a small hill, close to the apartment, where the girl disappeared. The costs of this obviously add up, and some of the police involved were raising questions over why so much money was being invested in the search for one girl. In this case, it was not just Internet trolls and misanthropes who were fuelling the debate, but the core concepts of cost and police priorities were being raised. This case reflects a wider and deeper meaning.

Before the demonisation of technology takes over, it needs acknowledging that technology enabled the Find Madeleine campaign at each anniversary to create an updated photo to match the passing of time. Technophobia and technophilia are different sides of the same coin. As with sex robots, this is a mirror that society can both gaze into to extrapolate an identity, or project its own fantasies. As we aged, Madeleine, in digitalised form, was aging with us, her photo a backdrop to our lives, appearing in a number of places, including Eddie Stobart lorries moving across the United Kingdom.

Digital images of the missing aging with us reflect the ability of technology to enhance the real, and raise a number of related questions. If we had a child that went missing, would we seek to build a Replicant, if the technology was available? If we had a partner who died tragically young, and it was possible to produce a copy via cloning or through robotics, would we wish for them to be still with us? Paradoxically, there is the ideologically enforced belief that when people go they go and we should not elongate this. Indeed, as medical advancements increase this is matched by the intense desire to allow euthanasia. Many stories, as we shall see, have developed the notion of a real child versus a 'false' child, Pinocchio being the paradigmatic example.

Those who resented the publicity the McCanns received made the negative point that one missing girl should not be prioritised over others, but the use of the media by the parents had always been a double-edged sword. The Portuguese police had requested the parents not go to the media. The parents believed that the first few hours following the disappearance were crucial, and that the police were not doing enough, so they ignored this request. According to the family, it was hoped that the fast-moving nature of the media, the use of technology across all platforms would help in raising the alarm, spread the word, and bring the girl back. But this was just one of many elements of the case that raised suspicion against the parents, especially by the Portuguese police, who should have been the first investigators as per protocol. The family's ability to be pro-active and technologically progressive and advanced in their search worked against them, as did their middle-class status and their wealth. Whether this hatred came from envy is difficult to tell, but they were resented for being able to express themselves by those who may often be on the fringes of establishment discourse. The latter's only means of attack, therefore, was to use technology against the couple and go trolling underground, and the movement against the McCanns grew. This can be viewed as a case of the 'haves' attacking the 'have nots', a mini class war, and it was covert and revealed a culture of envy and jealous, where intelligence is derided.

Only later when, for example, public figures like Stella Creasy decided to vote for the Conservative government in their bombing of Syria, did this form of electronic attack become more prolific. The relevance here is the suggestion that to some degree we are all now robots, having incorporated computer technology within our body, from our use of smart phones for example, that any trolling and cyber attacking is a physical attack. A comparison may not be enlightening, given that the McCanns were not paid public officials; they only made themselves public for the single purpose of gaining the return of their daughter. To their antagonists they were just as guilty, or even more so. Those who claim to have nothing to hide do not see any civil liberty issues at stake when it comes to surveillance. This suggests a certain blind trust in the benevolence of the state, which does not function alone globally, and now paradoxically the state is not publically accountable given the emphasis on private enterprise. While Ed Snowden and Julian Assange attack the state for prying into every private area, the majority seem perfectly happy to make their private life public, as McLuhan predicted, showing off every little detail of their lives on some form of social media.

Even popular fiction films, such as the James Bond film *Spectre* (Sam Mendes 2015), make the obvious point that once methods of surveillance are privatised then power is handed over to those supplying the surveillance. In the United Kingdom, the prisons and care homes have been privatised, making them less accountable. Paradoxically, while the public are increasingly observed and surveillance is part of everyday life, private industry creates a lack of transparency. The prisons, care homes, and schools that are taken out of the public system exist in their own private world, with little accountability. Robots are being produced to care for people, as we have seen, be this sexual or otherwise. Even surveillance is left to robots, and we are left with the uncanny notion of mass observation through CCTV by robots, who themselves are CCTV. There is a mechanisation that takes place here, given the profit motive that dominates this work, the vulnerable no more than units within the system. From the cradle to the grave, these units must consume and produce under observable mechanised conditions. Technology produces calculations looking at the cost effectiveness of those involved, regardless of the actual care provided. The exact point this happens without human input is an interesting question that is continually being argued over.

On a simpler level, scientists at MIT, such as Max Kanter and Kalyan Veeramachaneni, have developed what is known as the Data Science Machine, which can approximate human intuition when it comes to data. A central rejection of sex robots is that they will not be able to perform intuitively, but this raises the difficult question over what intuitive behaviour might actually be. We have already noted that there is no one definition of what a human is, what it means to be human, and when we try and define love it is too easy to get lost in clichés. It might be right wing and censorious to suggest that an absorption in pornography then dictates the fantasies and behaviours of those having sex, but Charlie Brooker's *Black Mirror* short film series delves into this issue astutely. The question is whether people really want to have sex with other apparently equal beings, or a reified fantasy which they can control. To put it bluntly, is virtual sex better than real sex? If we want everything controlled, if we are moving towards the latter, then the controllable sex robot is ideal, but the uncontrollable sex robot might be both more attractive, and more nightmarish, as we have seen.

This is equivalent to the 'Random setting' in Atwood's dystopia fiction. While her novel is interesting, it actually avoids the point. I have mentioned fears concerning robots relate to their uncanny resemblance to humans, but manufacturers of sex robots actually want them to be different, to be like dolls. Atwood explains that her fictional robots will be able to feel your skin; they will have senses:

> There are moving belts conveying thighs, hip joints, torsos; there are trays of hands, left and right. These body parts are man-made, they're not corpse portions, but nonetheless the effect is ghoulish. Squint and you're in a morgue, he thinks; or a slaughterhouse. Except there's no blood. (Atwood 2015, p. 187)

We have the paradox that their lack of humanity makes them horrific, as does their uncanny human resemblance, but as explained, in science fact the intention of these sex robots is not to make them resemble the human.

The relationships that result may indeed be similar to human relationships but they will be different. Stan in Atwood's sex robot factory, who is an imposter, thinks the only way he can fit in is to make jokes. This is, as we have seen, a primary human element. Again, as we have seen, language is of primary importance, although the men in the factory explain to Stan that the fact the sex robots have minimal language skills is an advantage. There is no nagging, no, 'have you taken the garbage out'. Stan, however, then misses Charmaine exactly because of her nagging, so what he hated is what he loved. He understands he will not get this from these sex robots. The wave of nostalgia that comes over him smells like orange juice, fireplaces, and leather slippers.

We began by considering the correlation between the human and the inhuman, the latter always remaining in the former, where the former's greatest desire is the latter. We do not have to be Slavoj Žižek to again acknowledge a Lacanian idea that it is the human who dwells in language. Why limit this to Lacan? All monotheistic religions basically say the same thing in their own way; in the beginning was the word. But Christianity takes it a step further, claiming the word dwelt with humans, and was in humans, the divine ghost in the machine, sublime machine in the ghost. This then re-animates the flesh. The difference between Heidegger and Lacan, is the former does not accept Being and logos as having this accord, and moves outside of this. The man who tested the robot may have ended up with a penis like a corkscrew, but it is language which is the torture-house for Lacan's version of psychoanalysis. Freud's psychopathology, that which makes us human, is actually the scars of this torture-house, from conversion-symptoms, inscribed in the body, up to total psychotic breakdowns (Žižek 2012). Lacan is committed to the linguistic and symbolic construction of desire, with his well-known view that the unconscious is structured like a language (Pope 2005).

As a reader, we know that even this is Stan's fantasy, raising the essential question: do we really know anyone? Is everything just a fantasy? At this stage in Atwood's fiction we might start to question whether Stan is actually a robot, especially if we are familiar with this genre. He has just been injected to look like he has died, his partner thinking she has actually killed him, and then he later is told he must dress up as an Elvis-bot to escape. His colleague Tyler explains to Stan that once the sex robots are assembled it is hard to know the difference, although if you see them assembled you always perceive them as an 'it'. This making and then marking them as an object, as an 'it', is obvious, as it is actually an object; the thing is a 'thing'. But when they start to move and talk we can question this 'it' status. There is a double nostalgia here because, even prior to the sex robot being animated, one of its creators regrets the way it will be seen as an 'it', the wider point being this is indeed how people treat each other anyway, especially women.

Once Stan, now renamed Waldo, is in the factory where they are making sex bots and meets the agent that might get him out of the place altogether, he learns that there is another level of 'robot' altogether. This is where they perform an operation on a 'real' person, to make them totally compliant to another. One can argue whether this is necessary, in reality, given the media do a very good job at this, in terms of the individual's relationship with the state. And then Stan/Waldo must keep his cool, as he comes across the realistic fake head of his partner. All of this takes place within a factory, and Atwood's stripped back prose does a good job in presenting a landscape of nothingness at the start of the novel, where there is a war between people for resources of the most basic kind. This is why a couple have chosen the so-called safety of living in a monitored and enclosed community, a form of human experiment, similar to Ballard's *High-Rise*.

Atwood is a well-known Canadian environmentalist, as well as a prolific novelist, and she spent her early childhood exploring nature with her scientist father and is an expert on nature. We have previously referred to Emerson and Bergson in the context of time, duration, and immortality, but it is not recognised enough that Emerson celebrated the new factory system. His point was that regions better suited for production will provide food, so he then has less of an obligation to sympathise for the hard-pressed Yankee farmers. His 1843 journal makes it clear that, 'Machinery and Transcendentalism agree well'. Unlike Blake, Dickens, and Carlyle, who predicted disasters of a new world full of dark satanic mills, Emerson promotes the American love of mechanisation. As Leo Marx has explained like, 'a divining rod, the machine will unearth the graces of landscape', for Emerson, who believed mechanical power would lead to access to the imagination, utopian, transcendent, value-creating faculty, Reason (Marx 1964). This is far from the warnings of Carlyle who wrote *Sartor Resartus* (1833–1834), and which Emerson read just before writing *Nature* and Melville read before writing *Moby-Dick*. The warnings of Atwood are predicted here by Carlyle.

To me [says Professor Teufelsdröckhu] the Universe was all void of Life, of Purpose, of Volition, even of Hostility: it was one huge, dead immeasurable Steam-engine, rolling on, in its dead indifference, to grind me limb from limb. (Marx, p. 179)

As with Frankenstein, we have the resurrection here of the notion of the severed limb, but in this context it is not to be gathered together and unified, to create a new being, symbolising hope. This is the antithesis; it is like the limbs in Stan's factory that Atwood equates with the slaughterhouse, a theme that returns in many journalistic accounts of visits to sex robot factories. There is an even greater warning in Carlyle, which is stark, and equates with the concerns already explored. What worries him is that our minds will become subject to the laws of matter and that, 'physical science will be built up on the ruins of our spiritual nature; that in our rage for machinery, we shall ourselves become machines' (Marx, p. 183). Conversely, Timothy Walker, a Harvard man, and others, found Carlyle's work blasphemous, because in their view this machinery was a revelation of a divine plan. Walker saw the ambition of democracy to free humans from bodily toil, but this has consequences. Think of Charlie Bucket's father in Roald Dahl's 1964 novel *Charlie and the Chocolate Factory*, who loses his job putting the caps on toothpaste tubes to a machine. Thankfully, for the kids and adults looking for a happy ending, he ends up getting a higher-paid job at the factory, fixing the machines that do his old job. Is there a parallel here? Atwood does her best to comment on the violence by prostitutes, who believe robots are making their labour defunct, and lowering prices, but in reality will prostitutes, who may lose their jobs, eventually get so-called better jobs, through fixing, coaching, and training sex robots?

Despite pornography having many advocates amongst women, and women being involved in the industry at every level, as directors and producers as well as the obvious actors, it needs acknowledging that the sex robot industry still seems male dominated. We have already seen this in terms of who are buying these products. Matt McMullen runs Abyss Creations' factory outside San Diego, which sells sexbot-style products for $5000 and up. This company was taken over by sexbot fan David Mills, the author of *Universe: The Thinking Person's Answer to Christian Fundamentalism*, a book praised by Richard Dawkins, and mentioned twice in his best-seller *The God Delusion*, and also commended by Carl Sagan's son Dorion.

For Mills, when he first ordered a doll, he felt they were extremely human-like, and his experience when his doll looked at him reminded him of the *Twilight Zone* episode, season 5 episode 3, originally broadcast on 11 October 1963, when William Shatner comes face-to-face with a monster on an airplane wing. For Mills, factory-built partners are at least as good as human partners. Is there something of the worship of technology here, a throwback to the nineteenth century previously touched on? There are, however, women involved in aspects of the industry. Annette Blair is the sales manager at Abyss Creations. It is acknowledged that for sex robot fetishists the pinball-like machinery is actually a turn on, the way exposed innards spin and light up, but these are not functional sex robots, they are a form of artwork. And taken in this direction, more women are involved, such as Stacy Leigh, a photographer of sex dolls, owner of nine, and an expert on them. Note the word sex dolls used here, not robots, but sometimes this is interchangeable, especially in popular culture, as we have seen. The dolls produced by McMullen have cropped up all over the place, including 20 television shows, such as *CSI: New York*, *My Names Is Earl*, *House*, and *2 Broke Girls*, and in 10 films, including *Surrogates* (Jonathan Mostow 2009), starring Bruce Willis hunting down killer androids and *2040* (Brad Armstrong 2015), when sex is outlawed and androids replace pornstars. In *Lars and the Real Girl* (Craig Gillespie 2007), a RealDoll stars opposite Ryan Gosling (Gurley 2015).

The paradoxes of the sex robots are contained both in *The Tempest* and the face of Hollywood stars like John Travolta, who have become fixed and plastic through surgery, a living doll. Stan in Atwood's novel worked in robotics, before giving up his life, and he knows how hard it is to construct facial expressions, such as smiles on robots. Fundamentally, the craft of acting involves empathy, stepping into another's shoes, but plastic surgery often makes natural expression difficult. For many Hollywood is still considered the epicentre of the global film industry and those acting within the industry, like Travolta, through surgery are destroying their ability to express themselves humanly. In our current brave new world, those with the wealth are creating a mini-society of plastic people. *The Tempest* is a prediction about the 'New World', written during the period concerning the European encounter with the Americas. Miranda, meeting strangers for the first time, magically falls in love.

Whether man, spirit, or robot, anything new is promoted as a transitional love-object, and of course Prospero her father is threatened. Ultimately, Prospero wants to control everything, through his magic, which we can read as science. Gonzalo, who lands on the island, represents the belief that there once was a golden age, and man was happiest without labour, which is a denial of history. Philosophically, Prospero is his antithesis, given he initially had to have his fate controlled by the elements and through reason and art he then dominated nature (Marx 1964). Robots represent Gonzalo's position of non-memory and where ultimately, for humans at least, labour maybe a thing of the past. The implanting of memories might be exactly what is required to humanise robots. The compliancy and peacefulness of robots are often seen as disturbing, as in the series *Humans*. Atwood deals with this with full use of paradox in *The Heart Goes Last* where the human Charmaine believes she has had an operation to make her besotted with Stan, which works, only to be told finally that this was a placebo, that the human mind is infinitely suggestible. This implies that psychology is stronger than medical science or robotics.

What needs to be understood fully is that it is the chaos and the mistakes of our wounded nature that makes us human that brings joy and creativity. Any desire for so-called higher perfect beings could be an attempt to circumvent any gaps in our own system, but if we fully filled these gaps there would be no life at all. This confirms the uncanny valley theory explained previously. The aim, therefore, should not be to reduplicate the human perfectly. We need to be more creative than this, despite any existential loneliness. This also has relevance in terms of disability studies and genetics, with the promotion of 'designer babies'. In the future, it will be robots seeking sex with humans, or at least comfort. The essential paradox over programmed randomness is difficult to comprehend.

Viewing the sex robot in female form as an attempt to circumvent the feminine and death, it might be viewed in its simplest form as a threat, because it leads to a stabilisation, challenging multifarious meanings, although this surely depends on the type of robot. 'Stable object, stable meanings: the surviving subject appropriates death's power in his monuments to the dead. A portrait may serve as just such a monument' (Ragland 1993). How about a robot? According to theologian Don Cupitt, morality is about giving up the old ideal of a life thoroughly examined and a unified life; in many ways it is about giving up memory and identity, as if stating one's difference invalidates the other. Freud's mission to turn neurotic misery into general despair has failed, and therefore an examination of the self should not take place. But, of course, in these terms, in this bland blanket postmodernism, this is all hypothetical anyway, given the nonexistence of the self (Lee 2009).

Our obsession with apocalyptic rhetoric has a long history. In the second century, *The Book of Revelation* was cited in writings more than any other book in *The New Testament* (Cohn 1999). Global terrorism, environmental catastrophe, predatory paedophiles, and now even sex robots form part of this rhetoric. Paradoxically, science fiction opens up alternative possibilities that can contribute to the reconstitution of authentic community. In the theology of many religions evil has or will be overcome because it is essentially not real. Postmodern interruptions have not moved beyond the Nietzschean view that truth is only appearance and anything that opposes this is illusory. In Dick's *Do Androids Dream of Electric Sheep?* only androids have false memory systems, yet random faults, be it in memory or otherwise, are that which differentiates the human from the machine traditionally. Imagination and creativity are formed through mistakes, but in George Orwell's dystopia land Oceania in *1984* there are no gaps. Part of the complexity touched on here is summarised in Hegel's approach, where 'transcendence *is* infinity, that is, the impossibility of encompassing or totalizing alterity' (Schroeder 1996). Memory then reinstalls the true meaning of forgetting, that is, significance itself.

Žižek argues there are three different versions of apocalypticism: Christian fundamentalism; New Age; and techno-digital-post-human. He is probably too narrow with this trilogy, but these three possess the connected view that humanity is approaching a zero-point of 'radical transmutation'. While referring to a wearable interface called SixthSense, developed by Pranav Mistry, of the Fluid Interfaces Group at the MIT Media Lab, he does seem excited; 'just think how such a device could transform sexual interaction'. The device merely aids seduction, giving you details about a potential sexual partner, such as how good they are at sex (Žižek 2011). Celebrating this device with enthusiasm, especially its low price, Žižek is writing in 2011. In the five years since, this has disappeared without trace.

In Stanley Kubrick's 1968 film *2001: A Space Odyssey*, HAL, the onboard computer who becomes the enemy of the humans on board the spaceship *Discovery I*, cries out in pain and fear when his circuits are finally taken apart. Notice, I use a gendered pronoun here, as does Dylan Evans when exploring the 'computer that cried' (Evans 2001). It is as if we want to give a gender to this machine, to allow it to have a personality and take it beyond a mere 'it'. Kubrick's film, released in the year before Americans landed on the moon, acknowledges fully how humans, computers and robots must successfully interrelate to advance humanity.

Evans makes the same point I am making about unpredictability. Focusing on robots and emotions, he is clear that the future looks positive, where humans and robots will have relationships, including love relationships. There is the development of self-evolving software, where computer scientists work to generate random sequences of instructions, and allow these mini-programs (known as genetic algorithms) to compete. The comparison to biological evolution is overt. Those that are the most successful reduplicate, with the copying process deliberately made imperfect, so error occurs allowing for mutant programs. 'If this process is repeated for many generations, the beneficial mutations accumulate, leading to exceptionally effective programs that no human could have designed' (Evans 2001).

Even sexy robots like Gemma Chan in the Channel 4 television series *Humans* are a canvas for projection, being imaginative constructs beyond fact. What is uncanny about these robots is their connection with ghosts. The ghost is not tied to an historical period, such as the Scottish manor, but is accentuated and accelerated by modern technology, inhabiting a phantom structure (Derrida 1989). They can be used to explore society's relationship with slavery, raising issues concerning subservience and class, as with *The Tempest*, *The War of the Worlds* and *The Heart Goes Last*. As with all 'others', any form of robot can be used to make the human feel superior. This feeling, however, masks a deeper fear of being replaced by a superior form which, when viewed cosmically and historically, is partially inevitable. These forms are unlikely to be similar to the basic sex robots currently available, which are simultaneously horrific and comic. Following a poststructuralist trajectory, if language is to be the focus for inspiration not experience, and it is language that speaks, not the writer or programmer, then theoretically anything is possible. Atwood's novel *The Heart Goes Last*, which emphasises the ambiguities of the reality of love over any pre-programmed love, explains that in many ways people want an excuse to be controlled by another. What this other might be should not prevent us turning away in horror, unless we choose to not confront ourselves.

References

Atwood, Margaret. 2015. *The Heart Goes Last*. London: Bloomsbury.
Bell, Aaron. 2011. The Dialectic of Anthropocentrism. In *Critical Theory and Animal Liberation*, ed. John Sanbonmatsu, 166. Plymouth: Roman & Littlefield.

Cohn, Norman. 1999. *Cosmos, Chaos & the World to Come: The Ancient Roots of Apocalyptic Faith*. New Haven, CT: Yale University Press, 212.

Demetriou, Danielle. 2015. 'My Weekend With Pepper, The Robot With Emotions'. *The Telegraph*, November 28. http://www.telegraph.co.uk/news/worldnews/asia/japan/12022795/My-weekend-with-Pepper-the-worlds-first-humanoid-robot-with-emotions.html. Accessed 01 April 2016.

Derrida, Jacques. 1989. '*The Ghost Dance: An Interview with Jacques Derrida*'. Trans. Jean-Luc Svobada, Public, no. 2, 69.

Edwards, Jason. 2005. *Eve Kosofsky Sedgwick*. London: Routledge.

Evans, Dylan. 2001. *Emotions*. Oxford: Oxford University Press, 99, 115.

Gurley, George. 2015. Is This the Dawn of the Sexbots? *Vanity Fair*, April 16. www.vanityfair.com/culture/2015/04/sexbots-realdoll-sex-toys. Accessed 02 April 2016.

Lee, Jason. 2009. *Celebrity, Pedophilia, and Ideology in American Culture*. New York: Cambria, 323.

Marx, Leo. 1964. *The Machine in the Garden. Technoloy and the Pastrol Ideal in America*. Oxford: Oxford University Press, 231–234.

Pope, Rob. 2005. *Creativity. Theory, History, Practice*. London: Routledge, 94, 95.

Radford, Tim. 2016. 'Touching Robots Can Arouse Humans, Study Finds'. *The Guardian*, April 5. www.guardian.com/technology/2016/apr/05/touching-robots-can-arouse-humans-study-finds?CMP=Share_AndroidApp_Messenger. Accessed 05 April 2016.

Ragland, Ellie. 1993. 'Lacan, the Death Drive, and the Dream of the Burning Child'. In *Death and Representation*, eds. Sarah Webster Goodwin and Elisabeth Bronfen, 14. London: The Johns Hopkins University Press.

Roszak, Theodor. 1995. *The Making of a Counter Culture: Reflections on the Technocratic Society and its Youthful Opposition*. Oakland: University of California Press.

Schroeder, Brian. 1996. *Altared Ground: Levinas, History and Violence*. London: Routeldge, 105. Italics in original.

Wiseman, Eva. 2015. 'Sex, Love and Robots: Is This the End of Intimacy?'. *The Guardian*, December 13. www.theguardian.com/technology/2015/dec/13/sex-love-and-robots-the-end-of-intimacy. Accessed 31 April 2016.

Žižek, Slavoj. 2011. *Living in the End of Times*. London: Verso, 337.

Žižek, Slavoj. 2012. 'Hegel versus Heidegger'. *e-flux*, #32. http://www.e-flux.com/journal/hegel-versus-heidegger/. Accessed 09 April 2016.

INDEX

© The Author(s) 2017
J. Lee, *Sex Robots*,
DOI 10.1007/978-3-319-49322-0